Living Farms

Living Farms
Encouraging Sustainable Smallholder Agriculture in Southern Africa

Martin Whiteside

earthscan
from Routledge

For
the smallholder farmers of southern Africa
and Liz, Anya, Naomi and Kieran

First published in the UK in 1998 by
Earthscan Publications Limited

This edition published 2013 by Earthscan

For a full list of publications please contact:

Earthscan
2 Park Square, Milton Park, Abingdon, Oxon OX14 4RN
Simultaneously published in the USA and Canada by Earthscan
711 Third Avenue, New York, NY 10017

Earthscan is an imprint of the Taylor & Francis Group, an informa business

A catalogue record for this book is available from the British Library

ISBN: 978-1-85383-590-2 (pbk)

Text photographs: © Martin Whiteside

Typesetting and page design by PCS Mapping & DTP, Newcastle upon Tyne
Cover design by Yvonne Booth
Cover photograph © Martin Whiteside

Contents

PART I INTRODUCTION

PART II RESOURCE-CONSERVING TECHNOLOGIES

PART V THE WAY FORWARD

List of Figures, Tables, Boxes and Photographs

FIGURES

TABLES

Boxes

PHOTOGRAPHS

Glossary and Abbreviations

ACHRM	African Centre for Holistic Resource Management
ADB	African Development Bank
ADMADE	Zambian CBNRM Programme
ADMARC	Malawian parastatal marketing organisation
ADRI	Agriculture and Rural Development Research Institute
AFC	Agricultural Finance Corporation (Zimbabwe)
AGENT	Agribusiness Entrepreneur Network and Training Development Programme
AgREN	Agricultural Research and Extension Network
ALDEP	Arable Lands Development Programme (Botswana)
ALIN	Arid Lands Network
AMC	area management committee
AN	Ammonium Nitrate
ARAP	Accelerated Rainfed Agricultural Programme (Botswana)
ARPT	Adaptive Research Planning Team (Zambia)
AZTREC	Association of Zimbabwean Environmental Conservationists
BCA	Botswana College of Agriculture
CA	communal area
CAMPFIRE	Communal Areas Management Programme for Indigenous Resources (Zimbabwe)
CASS	Centre of Applied Social Studies (University of Zimbabwe)
CBNRM	community-based natural resource management
CBO	community-based organisation
CDR	complex, diverse and risk prone
CFF	Central Finance Facility
cm	centimetre
CORDE	Cooperation for Research, Development and Education (Botswana)
CSC	Christian Service Committee (Malawi)
CSO	Central Statistical Office
D	Compound D fertilizer
D&D	diagnosis and design
DAPP	Development Aid from People to People (Monze)
DDF	District Development Fund
DEAP	District Environment Action Plan

DFID	Department for International Development (UK, previously ODA)
DFO	district forestry officer
DoA	Department of Agriculture
DR&SS	Department of Research and Specialist Services (Zimbabwe)
DVO	district veterinary officer
DVTCS	Department of Veterinary and Tsetse Control Services (Zambia)
EDA	Environment and Development Agency (South Africa)
EDC	Environment and Development Consultancy Ltd
ENDA	Environment and Development Action (Zimbabwe)
EU	European Union
EWs	extension workers
FAO	Food and Agriculture Organisation (of the United Nations)
FONSAG	Forum on Sustainable Agriculture (Botswana)
FSR	farming systems research
FSRE	farming systems research and extension
g	gramme
GDP	gross domestic product
GMB	Grain Marketing Board (Zimbabwe)
GRZ	Government of the Republic of Zambia.
GTZ	German Technical Cooperation Programme
ha	hectare
HDI	human development index (of UNDP)
HRM	holistic resource management
ICAZ	Institute of Cultural Affairs in Zambia
ICRAF	International Centre for Research in Agroforestry
ICRISAT	International Crop Research Institute for the Semi-Arid Tropics
IDRC	International Development Research Council (Canada)
IFAD	International Fund for Agricultural Development
IIED	International Institute for Environment and Development
ILEIA	Institute of Low External-Input Agriculture (The Netherlands)
IMF	International Monetary Fund
INR	Institute of Natural Resources
IRDNC	Integrated Rural Development and Nature Conservation (NGO)
IRDP	integrated rural development programme
ITDG	Intermediate Technology Development Group
ITK	indigenous technical knowledge
IUCN	International Union for the Conservation of Nature (World Conservation Union)
kg	kilogramme
km^2	square kilometre
Kw	Kwacha

L	Compound L fertilizer
LAPC	Land and Agricultural Policy Centre (South Africa)
LFCU	Likwama Farmers' Cooperative Union (Namibia)
LSU	large stock unit
LWF	Lutheran World Federation
MAFF	Ministry for Agriculture, Fisheries and Food (Zambia)
MAWRD	Ministry of Agriculture, Water and Rural Development (Namibia)
MET	Ministry of Environment and Tourism (Namibia)
mt	metric tonne
N	Nitrogen
NASCUZ	National Association of Credit Unions of Zimbabwe
NCP	Namwala Cattle Project (Namibia)
NEAP	National Environment Action Plan
NGO	non-governmental organisation
NRM	natural resource management
NRI	Natural Resources Institute (UK)
NRTSFA	Ndola Rural Technical and Smallholder Farmers' Association (Zambia)
ODA	Overseas Development Administration (of the UK – now DFID)
OPVs	open-pollinated varieties
PDT	Palapye Development Trust (Botswana)
PELUM	Participatory Environmental Landuse Management (Association)
PLA	participatory learning and action
PPP$	purchasing power parity in US$
PRA	participatory rural appraisal
RDSP	Rural Development Support Programme (Namibia)
RRA	rapid rural appraisal
SACA	Smallholder Agricultural Credit Administration (failed 1994) (Malawi)
SACU	South African Customs Union
SADC	Southern African Development Community
SAP	structural adjustment programme
SARDEP	Sustainable Animal and Range Development Programme (Namibia)
SDP	Smallholder Development Project
SHDF	Self-Help Development Foundation (Zimbabwe)
SSA	sub-Saharan Africa
T&V	training and visit extension system
UNDP	United Nations Development Programme
USAID	United States Agency for International Development
VDCs	village development committees (Malawi)
VIDCOs	village development committees (Zimbabwe)
VMC	village management committee

WTO	World Trade Organization
ZCF	Zambia Cooperative Federation
ZFU	Zimbabwe Farmers' Union
ZNFU	Zimbabwe National Farmers' Union

Acknowledgements

The author would like to thank the many farmers in Southern Africa, and those working with them, who gave up their time to discuss their livelihoods and work with the author, thereby providing most of the information and ideas for this book. They are too many to list by name, but their help and inspiration is gratefully acknowledged.

This research was funded by the Department for International Development (DFID) of the UK. However, the findings, interpretations and conclusions expressed in this book are entirely those of the author and should not be attributed to the DFID, which does not guarantee their accuracy and can accept no responsibility for any consequence of their use.

Many people contributed to the research and writing of various reports which led to this book, including Kagiso Baatshwana, Cindy Berman, Stephen Carr, Chileshe Chilangwa, James Copestake, John Erskine, Saliem Fakir, Steven Giddings, Jos Martens, Rob Mellors, Ketsile Molokomme, Noel Oettle and Piers Vigne. In addition, Chileshe Chilangwa, James Copestake, Stephen Carr, Jonathan Kydd, Richard Librock and John Wilson provided comments on the draft manuscript. Sue Gibbs provided invaluable administrative support and Kathy Benzinski subedited the manuscript.

The help of all these people and organisations is gratefully acknowledged; however, the author is responsible for the contents of this book and any errors or omissions contained within.

Preface

This book is about finding ways to encourage agriculture in Southern Africa to be both more sustainable and also to play a greater role in the eradication of poverty. It is easy to be unfairly critical about the state of agriculture in the region – currently agriculture is providing a major part of the livelihood for well over half the population of Southern Africa, and is managing to do this even as the population doubles every 25 or 30 years. This is a major success story, due largely to the skills and tenacity of the smallholder farmers in the region, as well as (and sometimes in spite of) ministries of agriculture and aid programmes.

Nevertheless, this success should not mask the fact that rural poverty remains acute and environmental capital is being eaten into; current practices are neither sustainable, nor are they ending poverty. The concentration by government and the commercial sector over many years on yield maximisation and the promotion of bought inputs such as fertilizer, coupled with a recent rush towards 'fiscal sustainability' (cutting subsidies on inputs and marketing), has caused a predictable crisis for those farmers unable to afford the inputs needed to make their current farming practice sustainable.

For a long time there has been insufficient emphasis on sustainability and those technologies using lower external inputs. The reasons for this are logical:

- Research and extension of low external input technologies tend to be long term and more complex, and therefore are often neglected.
- Most of the agricultural establishment has been trained within the yield maximising, high external input ethos.
- An increasing proportion of research and demonstration is being done by input suppliers, who naturally emphasise the use of bought inputs.

This means that alternatives to external input technologies have been seriously underdeveloped and underpromoted.

It is not suggested that using bought inputs is always wrong but rather that the emphasis has been seriously biased. Paradoxically, due to the very low use of external inputs in the region, there is generally scope for increased use of external inputs as well as a much greater emphasis on low external input technologies. It is not an either/or situation; both are needed to achieve sustainable intensification.

Developing appropriate technologies will have marginal effect, unless the local and national environment provides incentives for sustainability. The traditional environmental management practices of many communities have been undermined by:

- inappropriate state interference;
- the changing pressures on land and traditions; and
- the lack of clear natural resource property rights.

Communities need to be empowered to take a more effective role in sustainable natural resource management. This means working in new partnerships with government, the private sector and non-governmental organisations (NGOs). National policies need to be integrated into new thinking, arising from the community level.

Southern African smallholder farming needs to go through a transition, including recovering from decades of inappropriate policies and absorbing what (with the right policies) should be a temporary, very high rate of population growth. The transition to sustainability at the farm and community level needs to be supported, even if this means delaying sustainability at an agricultural services level; this may mean more long term donor funds or the cancellation of debts to reduce repayment levels. What is essential is that policies and programmes support a transition towards sustainability, rather than towards greater dependence. During the transition, the basic needs of the rural poor will need to be supported, and the challenge is to do this in a way which will not undermine progress towards sustainability.

Martin Whiteside
Gloucestershire
May 1998

Part I

Introduction

Chapter 1

Encouraging Sustainable Smallholder Agriculture in Southern Africa

WHO THIS BOOK IS FOR

This book is for all those people working on agriculture and rural development issues in Southern Africa. This involves individuals with many different skills working in many different ways, including:

- those working in community development;
- planners and policy-makers; and
- those working in agricultural research and extension.

Like the people working on rural development, the subject matter covered by this book is diverse, including technology, community development, agricultural service development and policy issues. This is intentional; many of us know a lot about our own specialist areas, but are less knowledgeable about the other specialisms that make up the whole. A multidisciplinary approach is essential for encouraging sustainable agriculture. This does not mean that we have to become specialists in everything, but it does mean understanding a certain amount about the different parts that make up the whole.

Providing examples of experience from South Africa, Namibia, Botswana, Zimbabwe, Zambia and Malawi, the book targets those people who want to learn from different experiences in these countries. This does not mean following blueprints developed by others, since programmes and policies need to be developed according to local circumstances, and in participation with local communities – but it does mean learning from the successes and failures experienced by others and adapting them to our local reality.

Although this book concentrates on smallholder agriculture, this does not imply that there are no problems of sustainability on large-scale farms – there are. However, it is believed that most of the potential readers are working with small-scale farmers and so this is where the focus is.

WHAT IS SUSTAINABLE SMALLHOLDER AGRICULTURE?

There are many different definitions of sustainability and sustainable agriculture; however, for the purpose of this book the following definition has been used.

> *Agriculture which meets today's livelihood needs, without preventing the needs of neighbours or future generations from being met; this is achieved by the continuous efforts of men, women and children to adapt complex rural livelihoods to a changing environment, so as to protect and enhance the stocks of natural, physical, human and social 'capital' available to themselves and to future generations.*

Some important points to note from this definition are:

- 'Meeting livelihood needs' implies agriculture contributing to the ending of poverty.
- 'Livelihood needs' goes beyond food security, to include the various components required for satisfactory living, including cash.
- The concern with neighbours means that it is not sufficient for one household to farm sustainably, if by their actions (for instance, using all the best land, or causing downstream pollution) they affect the sustainable livelihoods of others.
- It goes against the tendency to assume human omnipotency over natural resources.
- It allows for a limited substitution between the forms of capital – indeed the environment is a resource that is inevitably changed during the development process, but not necessarily destroyed (for example, converting 'bush' into fields is not necessarily unsustainable, it depends on the way the fields are farmed and whether sufficient bush is preserved to maintain other environmental balances);
- The 'changing environment' refers to the physical (rainfall), economic (prices) and social environment (attitudes of young people etc.);
- There is a recognition that the situation is not static – continual learning and changes are taking place.

There is a difference between sustainable agriculture, as defined above, and the sustainability of agricultural services, programmes and policies. The latter refers to the ability of a programme or services:

- to attract sufficient political support to secure permanent (usually government) funds and/or;
- to find ways of covering its costs (by levying charges), to keep going.

The sustainability of programmes and policies has become an important priority, particularly of some donors. However, it is argued that excessive attention to sustainable programmes and policies can sometimes undermine sustainable agriculture; ultimately no programme is sustainable if its users and clients are not sustainable (see Chapter 13).

Smallholders are a very difficult group to define, as they embrace a wide range of households, farm sizes and types of farming. Typically, they earn part or all of their livelihood from agriculture, cultivating up to about ten hectares (ha) (less where cultivation is intensive or irrigated) and may have a small herd of livestock, managed mainly by household labour.

Sustainable agriculture is not just about appropriate technologies being used by farmers. Often activities promoting sustainability, such as the management of common grazing, or collaboration over marketing, need to be undertaken at a community level. Therefore, community organisation is crucial. Other vital factors are the conditions in which farmers and communities are farming – secure access to suitable land, maintenance of roads, appropriate extension advice, markets, prices and many other components of the external environment can either help or hinder agricultural sustainability. These three aspects which contribute towards agricultural sustainability are shown diagramatically in Figure 1.1. This book is structured around these three conditions, which are necessary for sustainable agriculture.

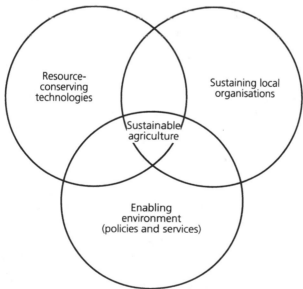

Source: adapted from Pretty, 1995

Figure 1.1: *Conditions for Sustainable Agriculture*

ARE CURRENT AGRICULTURAL PRACTICES SUSTAINABLE?

The answer to this question is not as straightforward as it seems at first. Both acute and long-term environmental problems exist in Southern Africa, some of which are linked to farming practices. However the overall situation can be difficult to assess since environmental degradation has sometimes been reported in an alarmist fashion by commentators who underestimate the resilience of many Southern African ecosystems. Some commentators try to draw attention to a particular issue of concern to them, in order to break through complacency, rather than present a balanced overall picture. Wiggins (1995) has pointed out that environmental degradation is less widespread than thought and, contrary to some expectations, the problems often diminish as population density increases.

Recent academic work in parts of West Africa is overturning the accepted academic view that deforestation on a large scale has been occurring everywhere as a result of smallholder agricultural practices over several generations. Instead, new evidence shows the reverse – smallholder agriculture has caused an increase in the number of trees (Fairhead and Leach, 1995; Fairhead and Leach, forthcoming). It would be wrong to jump to the conclusion that the same is true in Southern Africa, but it must lead us to treat the establishment view in Southern Africa with caution.

One influential report considered that 20 per cent of Southern African soils need some degree of rehabilitation, and the degradation and loss of productivity is continuing (SARDC/IUCN/SADC, 1994). Linking either measured or estimated soil loss to productivity decline can, however, be difficult – Scoones (1996) noted from Zimbabwe that, despite reports of soil losses over many years, the production of both crops and livestock for an area under study was maintained – or rose – over that period, with the main variability being due to rainfall. Sediment yield measurements from rivers often do not correspond to estimated soil loss, suggesting that in some cases soil is being redistributed rather than lost (Walling, 1984).

The population is growing through the region at a rate of about 3 per cent per year, with the different countries ranging from between 2.2 per cent and 3.8 per cent. At this rate, the region's population will double by the year 2018. SADC has stated that the increasing population is multiplying the effects of all environmental problems in the region (SADC ELMS, 1991).

There are, however, reasons to be cautious about applying an oversimplified approach of blaming population pressure for degradation:

- In some parts of the region, the problem is not so much high population density but high transaction costs, caused by poor transport and low population density, which are creating an unfavourable environment for sustainable smallholder intensification. Lack of labour is also a constraint in some areas.

- In Namibia, South Africa, Zimbabwe and (to a lesser extent) Malawi, the problem is not only lack of suitable land, but its unequal distribution.
- Smallholder yields tend to be very low in the region, compared with the potential; therefore, it is argued that the emphasis should be on developing the technologies and enabling environment to create sustainable intensification.
- How people use the land is crucial. Naturally, more people farming unsustainably will degrade resources faster; however, more people farming sustainably can also rehabilitate land faster. The experience from Machakos in Kenya (Tiffen et al. 1994) is an example of sustainable intensification, although the circumstances are quite specific (see Table 2.4).

The present rate of population growth (rather than the absolute population) does make the transformation to sustainability at the current time very difficult – production must rise by about 3 per cent per year just to maintain the current unacceptably low level of livelihood. There are indications, however, that the rate of growth of population is falling and that, as long as suitable policies and conditions are achieved, the rate will continue to fall to more sustainable levels (United Nations, 1996). However, even with a fall in growth rates, the number of additional people in the region will continue to rise for many years; the challenge is to support agricultural development that can both provide livelihood for the extra people and contribute to conditions which encourage lower population growth rates.

In moving away from an oversimplified Malthusian approach, it is important to avoid complacency. Although sustainable intensification could take place, this does not mean that it will take place. This depends on the combination of technologies, local organisation and enabling environment – the subject of this book.

The way people sustain their livelihoods is as important as their absolute numbers. The situation is also widely different in the various parts of Southern Africa.

Drier Areas

In drier areas, typically Botswana and Namibia, but also in parts of other countries, biological productivity is limited by rainfall and is therefore low. In many of these areas, low population densities mean crop agriculture tends to be extensive; often little or no fertilizer is used, and yields seem to have 'bottomed out' at a low but stable level – limited in most years by lack of rainfall. Human activity is causing some vegetative change, typically loss of tree cover, but in some cases there is bush-thickening due to grazing pressure and a change in heavily grazed areas from perennial to annual grass species. Although some consider this to be evidence of degradation and unsustainability, others point out that:

- Environmental change caused by natural variation in rainfall cycles is much greater than that caused by human activity, and the environments are naturally resilient.
- The total productivity (in economic terms) is continuing to rise in many areas that are considered degraded – perhaps a substitution of natural capital by human and physical capital is taking place.
- Many horror stories of desertification are based on observations at particular times of the year at particular places (such as near a borehole), rather than on a more balanced monitoring of the total environment over a longer period.
- Grazing intensification tends to be limited by periodic livestock losses in drought years.

It is unclear to what extent productivity can continue to rise, and whether a point of no return may be reached, perhaps leading to desertification. Better monitoring and a continued well-informed debate, involving both scientists and politicians, is needed.

In the drier areas (especially Botswana and Namibia) it seems unlikely that dryland crop agriculture alone will provide a viable route out of poverty; this is because low rainfall almost inevitably means that, in most years, production by poor households will be insufficient to provide a level of income above the poverty line. Livestock in these areas is potentially more lucrative. However, an increasing proportion of poor households have lost their cattle in recent droughts and this has only partly been compensated for by an increased number of goats. Therefore, although the large cattle owners, often with access to their own boreholes, support livelihoods well above the poverty line, the opportunity for poor households to become larger cattle owners and progress out of poverty by this route seems limited. For some people, alternatives, such as irrigation or chicken rearing, may provide adequate income, but the numbers are likely to be limited.

Less Dry Areas

In less dry areas in the region, soil fertility is more often a limiting factor in sustainable production. Traditionally, this has been overcome with shifting cultivation, however, increasing population densities in many areas are resulting in shorter fallow periods and even continuous cropping. This transition to continuous cropping is only sustainable if some means of replacing nutrients are used. However, rising fertilizer costs and low usage of organic methods of replenishing soil nutrients are leading to a crisis in some areas, such as southern Malawi. In areas where cultivation (or heavy grazing or clearing for wood) is taking place on sloping land, there is additional risk of erosion. In areas where this is occurring, the rate of soil loss is unsustainable and there is evidence of rapid and long term loss in

Table 1.1: *A Comparison of Basic Features between Densely Populated Rural Areas in Kenya and Malawi*

	Machakos (Kenya)	Zomba Thyolo and Blantyre Rural
Rural population	1.3 million	1.35 million
Total land area	1.36 million ha	0.632 million ha
Population density	97 persons/km²	214 persons/km²
Average cropped area per family	1.38 ha	0.805 ha
Per cent of area under non-basic food cash crop	25 per cent	<2 per cent
Per cent of land under maize	44 per cent	over 90 per cent
Total livestock units	360,000	approximately 80,000
Output in maize equivalents per head per year	700 kg (300 kg from actual food crops)	190 kg

Source: EDC, 1997b; Tiffen et al, 1994

productive potential. Areas at risk are increasing in zones of high landuse pressure, such as southern Malawi, parts of Zimbabwe and parts of South Africa.

The example of Machakos in Kenya is sometimes quoted as an illustration of sustainable intensification of agriculture, achieved alongside a rapidly increasing population. Although some comfort can be drawn from this example, it relied on quite specific conditions, such as the proximity to the market in Nairobi and the investment of salaries of urban workers in agriculture (partly for their retirement). Similar enabling conditions will be present in some areas of Southern Africa, but not all. In parts of southern Malawi, intense pressure on resources, greater than that experienced in Machakos (see Table 2.4), makes the switch to sustainable intensification particularly difficult and urgent. The need to make use of neighbouring areas with more natural resources (even cross-border to Mozambique), through trade or migration, is an option which needs serious consideration.

In high potential areas with adequate land, it is possible for agriculture to provide both sufficient food and sufficient income for poor households to rise above the local poverty datum line. However, in many of these areas this is not being achieved because of inadequate inputs and inadequate marketing opportunities. Sustainable agricultural techniques, improved services and local organisation will be required to achieve this.

Climate Change

All current thinking about sustainability could be upset by climate change. There are conflicting views on whether the climate has changed in Southern Africa or whether the recent droughts have just been a dry part of the long term cycle. Southern Africa's rainfall patterns have been highly variable for at least the last three centuries (Tyson, 1987) and there is evidence of an 18-year cycle of wet and dry years (Huntley et al, 1990). Although some climatologists consider that, so far, there is no detectable long term shift, there are indications of persistent desiccation with diminishing mean annual rainfall and increases in the severity of droughts (Magadza, 1994). Farmer interviews in different parts of the region show that many of them are convinced that the frequency of hot, dry periods has increased. A hot, dry period of several weeks at a critical time can be devastating, even if the total measured rainfall for the year appears adequate.

Much of the yearly variation in Southern African rainfall may be explained by changes in surface temperatures and currents of the Pacific, Atlantic and Indian Oceans (hence the El Niño effect). Understanding what causes these changes and predicting them over a longer period is fraught with difficulties. The ability to forecast the current year's weather came unstuck in 1997/98 in much of Southern Africa, with the predicted El Niño drought failing to materialise.

There is intense debate on the likely impact of global warming. It is thought that Southern Africa will be affected by global warming by experiencing increased aridity in the western parts and increased rainfall in the east. Overall, there are fears of increased desiccation and falling food security, due not only to reduced rainfall, but also to increased evaporation. It has been estimated for Zimbabwe that an increase in mean temperature of 2° Celsius would significantly reduce the food security in the dryer areas (Magadza, 1994). Sea level rise associated with global warming may also bring dangers of inundation and salinisation of productive low-lying river valley soils, particularly in Mozambique.

CONCLUSION

Achieving sustainable agriculture requires appropriate action at the technology, community, and policy and incentives levels. Sustainable agriculture needs to be differentiated from the fiscal sustainability of agricultural programmes; while both are important, one does not necessarily imply the other.

The question of to what extent present practices are sustainable is by no means clear cut and is the subject of much heated debate. There is evidence of vegetation change in all countries in the study, due to agricultural and other human activity. However, it is wrong simply to equate vegetation change with degradation or lack of sustainability. The most serious problems of sustainability are occurring in less dry areas with high population density and in areas where people are being forced to make use of steeply sloping land. Although widespread and permanent environmental damage is not yet evident in other areas, there is still a need to take precautionary measures to prevent problems in the future. Global warming could exacerbate the situation.

Rapid population growth means current production has to expand rapidly to meet livelihood needs. Although sustainable intensification could take place, this does not necessarily mean that it will take place – this will require appropriate combinations of technologies, community organisation and enabling environments.

Chapter 2

Agriculture in Southern Africa

SOUTHERN AFRICAN AGRICULTURE

Past and Present

The original farming system in much of Southern Africa was shifting-cultivation complemented, particularly in drier areas, by extensive livestock rearing. In the very driest areas, hunting and gathering have continued to predominate. However, the role of hunting, gathering and fishing as a complement to crops and livestock was, and continues to be, important in all areas, particularly in drought years.

Traditional shifting cultivation, with long fallows, could sustain up to eight persons per square kilometre (km^2); this density was reached in parts of Malawi, Swaziland and Zimbabwe at the end of the 1930s, necessitating shorter fallows and eventually continuous cultivation in some areas (Blackie, 1990). There has been a shift to maize cultivation during this century. This has caused concern about the increased susceptibility to drought of maize-based farming systems compared to millet- and sorghum-based systems. However the yield of maize per unit of labour (which is more important for many smallholders than yield per unit area), continues to make maize attractive (see Chapter 5).

It is tempting to assume a linear progression from shifting cultivation, through shorter fallows to continuous cropping. Although this may sometimes be the case, different systems are often practised at the same time and, over time, circumstances can encourage changes in the farming system in several different directions. Urban and peri-urban agriculture plays an important, but underrecognised, role in food production and support to household livelihoods. Policies and services are needed to make the most of what is likely to be a growing sector in the future. Water supply will probably be a crucial issue.

The vulnerability of the region to drought was highlighted by the 1991/2 drought, which resulted in a 60 per cent reduction in grain production across Southern Africa. There were 18 million entitlement losers, many of whom faced starvation and required famine relief. In addition to the immediate food insecurity, in a drought like this, many farmers lose their productive resource base, particularly livestock, which can take years to recover.

The relationship between the state and agriculture has changed markedly over the years. In the pre-colonial period, agriculture was a base of taxation for local rulers, and control over livestock, grazing and cultivating land was a major cause of various precolonial wars. In the colonial period, there was a degree of laissez-faire towards African agriculture and specific support for white settler farming. Where African farming was seen to compete with settler interests, by lower cost production or reducing the supply of cheap labour, vigorous policies were pursued to undermine African production or to force Africans into the labour pool (Parsons, 1982).

Effective organisation and lobbying by settler farmers resulted in favourable policies for the white large-scale farmers (for instance, in funding of agricultural research, which was orientated towards settler–farmer needs). Colonial administrations encouraged some technological innovations for smallholder farmers and were also active in veterinary control measures – perhaps because of the perceived danger of disease spreading from smallholders' to settlers' livestock. Anti-erosion practices were sometimes forced on smallholder farmers, causing widespread resentment, and memory of this can impede the introduction of erosion control programmes to this day. There were programmes in some countries, such as Zimbabwe, to introduce a class of African 'master farmer' who adopted certain modern techniques and, in return, gained an enhanced status and preferential access to land.

Post-independence policies towards agriculture have not been uniform across the region, although some trends can be identified. There were a number of smallholder-focused schemes, often integrated rural development programmes (IRDPs), with expenditure on rural roads and extension, including – partially planned and unplanned – growing subsidies to inputs and credit, often supplied through parastatals or state sponsored coops. However, in general the wider economic environment was not supportive of the smallholder sector in the 1970s and early 1980s (see Chapter 3). Inefficient state marketing organisations, unfavourable prices, inappropriate or non-existent technical support, and sometimes active discrimination against the smallholder sector in favour of state or collective farms (eg Mozambique), estates (eg Malawi), or white large-scale farmers (eg South Africa, Zimbabwe and Namibia) restrained the development of smallholder farmers. Where policy changed, such as in Zimbabwe after 1980, smallholder farmers responded rapidly, with expanded production.

From the mid 1980s onwards, while there has been a continued rhetorical commitment by governments and donors to smallholders, there has been, in practice, reduced outlays on smallholder support. The plan was to 'do more with less', achieving development through:

- greater efficiency of competitive market-driven service delivery;
- better, more focused remaining government services;
- more use of NGOs and greater diversity in service delivery patterns.

There are indications that the 'market liberalisation' era (1985–1997) is now giving way to a new paradigm of institutional development – with the state creating an enabling environment to foster institutional developments which reduce transaction costs (Jonathan Kydd, pers comm).

In the 1980s and 1990s, there has also been a growing recognition, at least among some of those involved in rural development, of the following:

- Smallholders have considerable skills very appropriate to their complex, diverse and risk prone (CDR) environment.
- There is a need to consider gender issues.
- There is a need for farmer participation.

Despite this, Southern African agriculture still suffers from too many generalisations. An example is the near universal condemnation of veld burning by ministries of agriculture and the environment throughout the region. However, many rural people continue to use fire as a management tool; are they always wrong? It is true that fire can be very damaging; but in some areas fire is a natural and necessary part of the environmental cycle, and in other circumstances, intentional, early, less hot fires can be a sustainable management tool which is less damaging than later season hot fires started accidentally or naturally (eg by lightning). The issue here is to understand the local complexity and diversity, rather than to make general assumptions – and to recognise that rural people often have good reason for their actions.

In several countries in the region, increases in life expectancy achieved in the last few decades have recently been reported to have worsened again. This is a result of both AIDS and in some cases worsening socio-economic conditions. The AIDS epidemic will continue to cause both massive human suffering into the next century and also severe economic disruption, with the loss of many people in the prime of their lives. For many, the family farming household will be the site for terminal care, emphasising both its importance and some of the strains it is likely to face in coming years.

Land Availability and Suitability

The suitability of different areas of Southern Africa for rainfed crop production varies enormously, with rainfall being the most important limiting factor. The population density in the individual countries of Southern Africa is very variable and can reflect historical factors as well as agricultural productivity. There are:

- the dry, sparsely populated countries of Botswana and Namibia (two per km²);
- the countries of Angola, Zambia and Mozambique, with more rainfall, but which are still generally sparsely populated (eight to 20 per km²);
- Zimbabwe and South Africa where there is moderate population density but where the very unequal distribution of land means there is acute land hunger (25 to 35 per km²);
- the higher population density countries of Swaziland, Lesotho and particularly Malawi (45 to 120 per km²), where land pressure is acute.

Even within those countries with low overall population densities, unequal distribution of land and people means that there are areas of acute land pressure. There is also a great range in the proportion of suitable crop land actually being used in different areas (see Table 2.1). Interpretation of figures can be difficult, since aggregate figures showing uncultivated suitable land do not necessarily mean that the land is available for smallholders to use. Unused suitable land may be remote or in private or state ownership. Some land listed as uncultivated may also be in long term fallow.

Deforestation is worse in countries with acute land pressure and is particularly severe in Malawi; however, what is defined as forest in Table 2.1 covers quite a wide variety of woodland types and therefore is not necessarily comparable between countries. The relatively high figure of deforestation in Mozambique perhaps reflects the war years, and a more recent longer term study indicates a less critical situation at a national level, but with considerable deforestation around major cities and transport routes (Saket, 1994).

Some areas of Southern Africa potentially suitable for grazing and cultivation are home to the tsetse fly. Control and eradication programmes are underway in a number of countries – some see this as an opportunity to increase the quantity of usable land; others are concerned about the environmental consequences of increasing human activity and cattle grazing in these areas.

The agricultural sector of most countries in Southern Africa remains strongly dualistic, with a relatively small number of large commercial farms and a large number of diverse smallholder farms. Typically, the commercial farms occupy the most favourable areas, are mechanised, use the latest scientific methods, and use moderate to high levels of inputs.

About 50 to 80 per cent of the population in the various countries are engaged in farming. Smallholder cultivated areas tend to be small, often

Table 2.1: *Land Areas, Population Density, Arable Land, Irrigated Land, Forest Area and Deforestation in Southern Africa*

Country	Land Area >000 km²	Population Density* (people per km²)	In use for crops (% of suitable)	Forest and Woodland (% of land area)	Annual Deforest- ation (1981–89)	Irrigated Land (% of arable)
Angola	1,250	8	5%	42%	0.2%	2.5%
Botswana	582	2	20%	46%	0.1%	0.5%
Lesotho	30	61	32%	10%	–	1%
Malawi	94	116	45%	31%	3.5%	1.7%
Mozambique	802	19	11%	18%	0.8%	4%
Namibia	824	2	–	22%	0.2%	<1%
South Africa	1,220	33	–	7%	–	10%
Swaziland	17	47	23%	7%	–	36%
Zambia	753	11	9%	38%	0.2%	1%
Zimbabwe	391	27	19%	22%	0.4%	7%

Source: adapted from UNDP, 1996; Alexandratos/FAO, 1995
* The average figure can be misleading, with much higher densities in some communal areas; for instance, smallholder areas in Malawi average 250 people per km² (EDC, 1997b).

0.5 to two hectares if hoe cultivation is used and one to ten hectares if animal traction or hired tractor ploughing is used. Smallholder grain yields tend to be low, typically 200 to 900 kg per ha in drier areas and 700 to 1500 kg per ha in wetter areas. Intercropping of a cereal staple with legumes and cucurbits is common, and in some areas root crops, particularly cassava and sweet potato, are very important, as are cash crops such as cotton and tobacco. The designation 'cash crop' is not particularly helpful, however, since 'food crops' like maize and beans are also often grown by smallholders for sale.

Despite the general increase in yields during the last 25 years as shown in Table 2.1, there is now an indication that yields in some areas are falling, with diminishing soil fertility the main factor. Much of the overall increase in production during the last 25 years has been due to an increase in the area cropped; in some areas there is sufficient land for this clearing of new land to continue, but in others, the only way to increase overall production is likely to be intensification.

Livestock are important throughout Southern Africa, though with quite marked differences between drier and wetter areas (see Table 2.2). There are, however, some important changes taking place:

- *The poor do not have cattle.* Even in drier areas where livestock farming is a traditional activity, the poor may not have livestock,

Table 2.2 *Population and Agricultural Production in Southern Africa*

Country	Population in Year 2000 (millions)	Percentage Population in Agriculture (Year 2000)	Average Annual Food Production (growth per capita 1979–92)	Average Yield of Principle Cereal (1989/91) kg/ha)	Percentage Change in Average Yield Principal Cereal (1969/71– 89/91)	Percentage Change in Area Under Cereal Cultivation (1969/71– 1989/91)
Angola	13	66%		maize 300	−66%	+39%
Botswana	1.8	55%	−3.1%	sorghum 210	no change	−2%
Lesotho	2.4	71%	−2.2%	maize 970	+48%	−35%
Malawi	12	80%	−5%	maize 1,200	−11%	+21%
Mozambique	20	78%	−2.1%	maize 370	−63%	+120%
Namibia	2.4	28%	−2.5%	millet 600	+33%	+43%
South Africa			−2.1%			
Swaziland	1.1	58%		maize 1,400	+60%	+1%
Zambia	12	65%	-0.8%	maize 1,700	+110%	−24%
Zimbabwe	13	63%	−3.3%	maize 1,800	+12%	−3%
SSA	673	66%				+15%

Source: adapted from UNDP, 1994; Alexandratos/FAO, 1995

particularly cattle. Poor households have been hard hit in recent droughts and the proportion without cattle has risen. For instance, in Botswana, 32 per cent of households had no cattle in 1981, but by 1996 this had risen to 49 per cent. Interestingly, despite comprehensive crop packages to support poor smallholders, the Botswana government has done little to help poor farmers restock (except for draught power); perhaps this reflects an implicit policy that cattle ownership among the poor should not be encouraged.

- *More goats, less cattle.* Goats (and to a lesser extent sheep) have increased relative to cattle in Botswana, Zambia, Zimbabwe and Malawi. While in Botswana in 1980 there were five cattle to each goat, in 1993 goats outnumbered cattle. In Zambia there has been a serious fall in the national cattle herd from 1,876,000 in 1983, to just 787,000 in 1993, and draught animals have fallen from 360,000 in 1990/91 to 190,000 in 1993/94 (GRZ/CSO, 1991; 1994). The reasons for this are: drought, with cattle mortality higher and goat number recovery being more rapid; and disease, particularly corridor disease in Zambia and contagious bovine pleural pneumonia in Botswana. There is some suggestion from Zambia that increased disease has been due to reduced veterinary services because of budget cuts.

Table 2.3: *Livestock in Southern Africa*

Country	Livestock Millions (1989/90 unless stated)*	Percentage of land as pasture
Angola	3.1 cattle 1.3 sheep and goats	25%
Botswana	1.6 cattle (1993) 2.1 sheep and goats (1993)	75%
Lesotho	0.5 cattle 2.5 sheep and goats	80%
Malawi	0.8 cattle (1995**) 1.1 sheep and goats	50%
Mozambique	1.4 cattle 0.5 sheep and goats	55%
Namibia	2.1 cattle 9.1 sheep and goats	not available
South Africa	11.9 cattle 37 sheep and goats	66%
Swaziland	0.7 cattle 0.4 sheep and goats	80%
Zambia	0.79 cattle (1993)*** 0.6 sheep and goats (1989/90)	65%
Zimbabwe****	3.5 cattle (1996) 2.9 sheep and goats (1996)	60%

Sources: World Resources Institute, 1992; various government reports
* In many countries, the number of cattle has fallen during the 1990s, due to drought; households holding relatively small numbers have often been the worst affected.
** The number of cattle has been declining in recent years.
*** The number has fallen rapidly since 1990 due to drought, east coast fever and corridor disease.
**** Numbers from communal and resettlement sectors.

- *Spread in draught power.* There seems to be a spread of cattle, and the use of animal draught, into areas previously without cattle which had a tradition of hand cultivation. Tsetse control has also facilitated the spread of cattle in some areas. Donkey draught is also a new phenomenon in some regions.

CHANGING APPROACHES TO AGRICULTURAL DEVELOPMENT

There has been a progression of approaches to improving smallholder agricultural performance in Southern Africa, which has broadly followed the development of thinking in the rest of the world.

Transfer of Technology Approach

This approach is founded on the belief that Western scientific methods are the key to progress, and that technologies successful in research stations will be successful with smallholder farmers. This was the basis of the Green Revolution of the 1960s and 1970s, in which fertilizer-responsive varieties, combined with organic fertilizer, increased yields markedly in relatively uniform and favourable areas of Asia and Latin America; however, they have so far had little overall impact in Africa. A home-grown Green Revolution occurred among Rhodesian large-scale farmers in the 1960s, with the release of the locally developed hybrid maize variety SR–52; this was followed by uptake of hybrid varieties by Zimbabwean smallholders in the 1980s. However, this second Green Revolution has proved to have limited sustainability. Most of those benefiting were found to be the better-resourced smallholders, and the state has not been able to provide the favourable marketing and other support, which encouraged this increased production, in the face of wider economic crises (Eicher, 1995).

Farming System Approach

The failure of the Green Revolution to have a significant impact in much of Africa (and also in less uniform parts of Asia and Latin America) led to an appreciation of the diversity and complexity of the smallholder farming system. Rather than concentrate on optimising the yield of a single crop, the wider agro-ecological context was considered; this became known as farming system research and extension (FSRE).

Farming system research generally retained a belief in the importance of Western scientific methods, but recognised that this had to be applied to a complex situation – with a mixture of crops being grown in each field, a mixture of micro environments being exploited by each farmer, a mixture of enterprises (both on and off-farm), making up household livelihoods, coupled with high climatic variability.

Training and Visit

In the early 1980s, at about the same time as the US Agency for International Development (USAID) was introducing farming systems research (FSR), the World Bank brought the training and visit (T&V) system of extension to Africa, following its perceived success in Asia. Variants of the T&V system have been adopted by many of the countries in the region.

T&V is a management system to improve control of conventional research and extension (Woodhouse, 1991). The introduction of T&V and FSR should have been linked, but generally they were not (Moris, 1991). T&V is based on:

- the functional unity of research and extension;
- clear lines of command;
- agreed messages;
- a set programme of visits;
- demonstration plots by farmers;
- two-way contact between 'contact farmers', extension workers and researchers;
- the division between subject matter specialist and extensionist;
- parallel systems of monitoring and evaluation.

The mode of operation of T&V in Southern Africa has been extensively criticised. T&V requires considerable resources which are often not available after the initial introductory period, and have become even less available in the current climate of structural adjustment. It is also very difficult to make T&V participatory, and many systems have operated on a top-down transfer of technology approach, despite the rhetoric. On the other hand, many extension offices, finding the T&V system too inflexible, are operating hybrid systems.

Farmer-First Approach

There has been a growing realisation that technological packages alone cannot remove the constraints facing small farmers. The farmer-first approach established the need to put the farmer at the centre of moves to develop sustainable agriculture in the 1990s. It involves a number of key concepts:

- Farmers, and farming communities, are knowledgeable – especially about local conditions, which is where outsiders' knowledge is weakest (Chambers, 1994a). This knowledge is evolutionary and involves both indigenous technical knowledge (ITK) and more recent knowledge obtained from a number of sources.
- Farmers are largely rational, responding in their own best interests to the diverse physical, economic and social environments in which they operate. However, rationality is exercised within a specific social and cultural context.
- Farmers need an enabling environment, including resources, information, security and power.
- Participation by farmers in the process of identifying and overcoming problems is essential.
- For scientists, as well as politicians, this new agenda can be very threatening and, therefore, barriers to its adoption can be assumed.

Various terminologies have been used to describe the process, including 'Farmer back to farmer' (Rhoades and Booth, 1982), 'Farmer first and last' (Chambers and Ghildyal, 1985), 'Farmer participatory research'

(Farrington and Martin, 1987) and 'Participatory technology development' (ILEIA, 1989). A range of techniques and approaches have been developed for working in a more participatory way with farmers, including rapid rural appraisal (RRA), diagnosis and design (D&D), participatory rural appraisal (PRA) and participatory learning and action (PLA).

Overlapping Approaches

The progression of approaches described above, from transfer of technology, through farming systems to farmer first has occurred to some extent in Southern Africa but, like elsewhere, the stages are not discrete. In practice, different approaches are being used at the same time; sometimes they complement each other, sometimes they conflict.

Most of the research establishments in Southern Africa, and the scientists working in them, are founded on the transfer of technology approach. Most have adopted some elements of the farming systems approach – often to a greater extent in rhetoric than in reality. In many cases, a farming systems component has been added on to the existing research and extension programme, rather than forming an undergirding philosophy and approach. Single-crop research programmes exist throughout the region in relative isolation from more integrated work also going on, often within the same institution. Although many organisations are using participatory appraisal techniques, this is not always linked to a farmer first or a participatory approach to the rest of their work (see Chapter 6).

In reality, a combination of approaches is probably inevitable and may, indeed, be effective. Often, the best progress is made when the diverse skills and experience of farmers and scientists come together to create a whole greater than the sum of the parts. However, this requires a much greater emphasis on sustainability and the poorest farmers than is currently the case (see Chapter 4).

COMPLEX, DIVERSE AND RISK-PRONE FARMING

Smallholder farms in Southern Africa can be described as complex, diverse and risk prone (CDR) (Chambers et al, 1989). Farming under these conditions, and making the best use of changing and variable circumstances, has been termed opportunistic farming since it does not follow a set pattern or blueprint (Scoones, 1996).

Complexity

Farms tend to be complex in a number of ways. Within fields, intercropping is common, which reduces the relevance of much monoculture research being done in research stations. Interaction between different components on the same farm is an additional complexity which can produce benefits (for example, cattle benefit crop farming by ploughing and providing manure, but also profit from the stover), but it is difficult to produce recommendations for either component in isolation. The labour needed at certain times for the different farm activities is the subject of a complex juggling act designed to produce the greatest overall benefit, even though it might not be possible to undertake each operation at the ideal time for the individual component. For instance, although early ploughing has been recommended in Botswana for 20 years, resource-poor farmers have not followed this advice, mainly because they do not have fit draught power available at this time of the year.

Farmers grow a large variety of crops, through intercropping, planting in sequence and by using different micro environments. Many so-called minor crops such as sweet potatoes, which do not receive much attention in national statistics or research and extension efforts, fill an important niche within the household livelihood. Farmers sometimes vary planting density within a field, making use of local micro environments, planting patches with slightly heavier texture more densely, or with specific inter-crops. Such complexity exploits the environment to the full, reduces risk and increases variety in the diet.

The farm is a provider for the changing needs of the household in the absence of banking and credit facilities. Thus sales may be made according to current necessity – a chicken to buy medicine, a goat to pay school fees, or a cow for a wedding. This is often perfectly rational behaviour in terms of the household as a whole, even though such sales do not always coincide with the optimum sale time when looked at from a profit-maximising perspective for the individual enterprise.

Diversity

Rural incomes are highly skewed throughout the region, even within the smallholder sector. Jackson and Collier (1988) found for communal areas in Zimbabwe that the top 10 per cent of households controlled 42 per cent of income, while the bottom 25 per cent controlled only 4 per cent. This skewed income was thought to be particularly pronounced in drier areas. In some parts of Southern Africa households tend to go through a cycle – starting with few assets, building them up through middle age and then selling them to provide food and other resources in older age.

Even between households at similar socio-economic levels, there may be very different ways of making up livelihoods. Therefore the balance of crops, livestock, off-farm employment, and hunting and fishing may be very different between neighbours. An external event or change in service provision can affect seemingly similar households in very different ways.

Farmers vary in their expertise and commitment to farming; in each local area some individuals are renowned as 'good' or 'keen' farmers, some as 'bad' or 'disinterested' farmers and the majority somewhere in between (the transfer-of-technology extension approach uses the rather patronising labels of 'innovators', 'adopters' or 'laggards'). There are expectations that 'all farmers need to be brought up to the standards of the best'. This misses the point that there are reasons for the diversity and is about as realistic as expecting everyone to run a four-minute mile. There is, however, considerable scope for interhousehold learning from the best practices already being used in the area.

Many agricultural programmes have not taken sufficient notice of the diversity among farmers. Worse still, agricultural policy, research and extension have often been directed at an unrepresentative sector of farmers – typically richer, male and less remote. Conditions in most experimental stations relate more closely to the conditions experienced by richer families; therefore, not surprisingly, the technology packages developed are often more suited to them.

Policy is usually disproportionately influenced by those who are politically powerful and articulate. In Botswana, the needs of the larger cattle owners, who are well represented among politicians and senior civil servants, have been well served by government policies in comparison to the large number of resource-poor farmers. Extension services often

Table 2.4: *Socio-Economic Differences Between Those with Draught Access and Those Without*

Dimension	Households with Draught Access (N = 80)	Households without Draught Access: fewer than two adult cattle or donkeys (N = 66)
Field area (ha)	6.2	4.5
Cattle	6.1	0.4
Donkeys	1.9	0.05
Goats	8.9	2.36
Ploughs	100%	38%
Grain output (bags):		
1990	19.2	7.3
1989	19.2	7.4
1988	26.1	14.9
Maize sales (bags)		
1990	2.9	0.8
1991	2.5	0.6
1988	7.6	1.6
Household size	8.3	5.9
Remittance access	72%	44%

Source: adapted from Scoones, 1996

Box 2.1: Gender and Sustainable Agriculture

Women remain the key players in most smallholder agriculture throughout the region. What this means in practice, however, can be very variable. Family structures are changing rapidly, with increased divorce and breakdown of the extended family (lessening both the security and constraints it provided).Gender roles vary from household to household, from area to area, and are changing over time. In very general terms:

- Inheritance patterns vary throughout the region, being patrilineal or matrilineal in different areas; there is some indication of an increased tendency towards patrilineal inheritance in some areas, and more even sharing of inheritance between sons and daughters in others. The degree of security a woman has over land and other farming assets, particularly on divorce or widowhood, can depend on the interaction between the individual family, local traditions (particularly whether patrilineal or matrilineal) and national law (which is slowly becoming less gender biased in most countries).
- Husbands and wives may have separate fields or share the same fields; even when they have separate fields, they may both work on each other's fields, and the produce may be kept separate or combined.
- Women tend to be responsible for food crops, storage and processing; however, men may help or be responsible for various stages. There is great variability over who decides key issues, such as the proportion of food versus cash crops to plant, or how much grain to sell and how much to store, or on making investments such as buying a plough or planting trees. In some cases, although a woman may do all the day-to-day work, she may not be able to take larger decisions – which can constrain timely management if the husband is away working, for instance.
- Although men are often considered necessary for ploughing when using animals, in nearly all areas there are examples of individual women ploughing; this is particularly common when donkeys are used.
- Individual animals are often owned by different members of the household, both men and women – although the animals are often managed together in the same herd, with either men or women doing the work. Although small-stock (chickens, ducks, goats, sheep and pigs) tend to be more often the responsibility of women, there are exceptions. Men tend to be responsible for issues relating to common property grazing management and livestock water points, although again there are exceptions.
- Women (and children) tend to be responsible for gathering firewood, although when firewood procurement and sale become a commercial activity, then men are generally involved. Women (and children) tend to be responsible for gathering wild fruits and vegetables. Both women and men may be responsible for planting or protecting trees; men are sometimes considered necessary for the heavy work of clearing and destumping bush to make fields – although there are examples of women taking this on.
- Though in most communities men tend to be overrepresented in rural community decision-making structures, there is often some involvement of women. Much less is known about the actual role of men and women in informal decision-making processes. The age divide in decision-making may be even more marked than the gender divide.

The increasing numbers of female-headed households throughout the region have been noted in many reports (amounting to 43 per cent of households in rural Botswana), along with concern about the high degree of poverty found in some of these households. The households tend to have high dependency ratios, low labour availability and low access to draught power and cattle. However, female-headed households are very diverse and generalisations need to be treated with caution – households where a man is still nominally the head, but is away working, are often different from those where the woman has sole responsibility; indeed, in some areas of rural Zimbabwe, female-managed households (where the man is away working, but sends back remittances) are among the best off.

Women and men are treated differently at three different levels; some of these may be in contradiction to each other:

- in national law;
- in traditional law;
- in existing social practice.

For instance, in Botswana, national law has been restrictive towards women in agriculture because of its effect on land rights and credit. A process is underway to end this by changing the Deeds Registry Act and the Marriage Act. Women are also disadvantaged under customary law over land and property rights, which affect inheritance and rights on divorce. The majority of the population follows patrilineal inheritance patterns, and women often need support from a male relative in order to get access to land (EDC, 1997(a)).

Customary law and social practice change over time. For instance, traditionally among the Tswana the eldest son inherits most of the cattle and other belongings from his father, with the widow and daughters getting little or none (the son would, however, also inherit some obligations to care for them). Recent trends have been towards the widow inheriting most of the cattle and belongings (even if her son looks after the cattle) and for a more equal division between all children.

Women and men tend to be treated differently by government and NGO agricultural programmes:

- The majority of programme managers, fieldworkers and researchers are men; despite some initiatives in gender training this is likely to have an impact on how the programme relates to women farmers.
- Although there is more awareness of gender, it is often an issue that is tacked on to agricultural programmes, rather than the whole programme being based on a thorough gender analysis – including the different roles, perceptions, constraints and ways of working with women and men in agriculture.
- Some programmes, such as the Arable Lands Development Programme (ALDEP) (see Box 11.12), have tried to improve the participation of women by affirmative action and offering better terms to female-headed households; however, this does not necessarily mean that the overall design of the programme has been based on a thorough gender analysis.
- There are a small but growing number of programmes working specifically with women farmers – the challenge is to ensure that some of the lessons they are learning, and approaches they are developing, are transferred to more widespread programmes and replicated more extensively. There are also a growing number of organisations with a specific expertise (such as women and the law, or gender training) which should be of use to other organisations involved in sustainable agricultural development.

Of importance for the future are the differing aspirations of boys and girls, in relation to farming. There are indications that these aspirations are quite negative, which is an issue which should be looked at further since it is unlikely that men or women will become skilled farmers, with a long term perspective in nurturing the environment, unless they feel positive about being a farmer.

mainly address men, even where women do most of the agricultural work and take most of the farming decisions. The extension services, by working through contact farmers or active members of farmers' groups, tend to concentrate their efforts on the richer and middle income farmers, who are often more self-confident and demand more attention. The whole range of services provided, including sale of inputs, credit, grants, marketing, and veterinary support are not only often inappropriate for resource-poor farmers, they may also be completely unavailable away from the main roads, and therefore exclude whole groups of farmers.

Differences in wealth are usually reflected across a number of key variables between households, and one often reinforces others. Unsurprisingly, in Zimbabwe, households with access to draught power are also better off in a number of other ways as well (see Table 2.4).

Issues such as gender and age can add to diversity, both within a household and between households. Within a household, different generations of women and men may have different roles, skills and priorities. Households headed by women or older people can be different from those headed by men or younger people. However, it is dangerous to either generalise or assume that such differences are static (Box 2.1).

Risk

Farmers in Southern Africa face a wide variety of risks – weather, war, robbery, pests and diseases, and price changes. Drought has been a recurring feature since agriculture began in Southern Africa, but the 1980s and 1990s saw particularly severe droughts. Staggered planting dates, growing a mixture of early- and late-maturing varieties, farming a number of individual fields spread over a wide area, and spreading livestock between different kraals of different family members are all techniques used to reduce risk.

The effect of illness or death within a household is an additional risk for poor households who are dependent on adult labour resources for survival. Illness was found to be a major reason for some Mozambican returnee families failing to reestablish livelihoods on their return after the end of the war (Whiteside, 1996a). The risks faced by smaller farmers as a result of increased robbery in the region are possibly an as-yet underreported constraint, particularly in terms of livestock keeping nearer towns.

Naturally, the high risks and lack of reserves make Southern African smallholder farmers nervous of taking further risks, like resource-poor farmers anywhere. Sometimes the risk-adverse nature of smallholder farmers is wrongly used to assume that these farmers will be slow or reluctant to try new approaches. In fact, when smallholder farmers perceive a new opportunity, and judge the risk-benefit ratio to be acceptable, they do respond rapidly. This was seen in the way Zimbabwean farmers responded to the new conditions after independence. Another example is the way in which Malawian farmers have taken up growing chillies for export, often using small patches of ground and therefore minimising their opportunity costs and the risks (Schwartz, 1994).

Risk is a concern for policy-makers because:

- Risk management strategies need to be understood before 'improvements' to the farming system are suggested.
- Underinvestment in technology or management may be a result of being unwilling or unable to take risks.
- Risk and the resulting shocks may undermine sustainable rural livelihoods.

Recognising that the world is an uncertain place, and in particular that agriculture is especially uncertain, has implications for policy-makers. Instead of fixed responses, an adaptive management and policy approach is needed in which the situation is monitored and changes made accordingly. This means acknowledging that everything is not known at the outset – a concept difficult for politician and senior administrator alike.

CONCLUSION

Southern African agriculture is varied, depending on history, rainfall distribution and differing population densities. Recent severe droughts have undermined smallholder resources and left many of the poorer households without cattle – which has had a knock-on effect on draught power, land cultivation and fertility maintenance.

Policies for smallholder support have been largely founded on the transfer of technology approach, although more holistic and participatory approaches have also been incorporated.

The complexity, diversity and risk-prone nature of smallholder agriculture has generally been underestimated by agricultural professionals. Research, extension and the provision of services for highly complex and very diverse farming systems is difficult if traditional top-down approaches are used, in which complexity and diversity are seen as problems. Local farmers instead use the complexity and diversity as part of the solution. Agricultural service providers need to work with local farmers and communities to make the best use of this variability. Services and policies must support farmers in reducing their risks.

Chapter 3

Can Sustainable Agriculture End Rural Poverty?

RURAL POVERTY

In some countries in Southern Africa, and among some donors, there has been a shift in concern from rural poverty to urban poverty. There are a number of reasons for this:

- Urban poverty is increasing and structural adjustment programmes have made urban living very difficult for many households.
- Urban poverty tends to be more visible – with slums, beggars, increased crime and riots.

However, statistics and surveys throughout Southern Africa show that the incidence of poverty is higher in rural areas, the absolute number of poor people is also higher, and extreme poverty is concentrated in rural areas. Therefore, while concern for poverty in urban areas is necessary, this should not be allowed to detract from the priority of addressing rural poverty.

It is also important to remember that urban and rural poverty are not separate – many households have members working in urban areas and other members farming in rural areas. Food and money pass in both directions, depending on the time of year and the success of the harvest; having household members in both urban and rural areas is an important risk-reduction strategy. Many urban dwellers take leave at ploughing or harvest time, providing extra labour but also having a low-cost 'holiday'.

There is considerable difference in wealth between different countries in the region, both measured by gross domestic product (GDP) and the United Nations Development Programme's (UNDP's) human development index

Table 3.1 *Poverty Indicators in Southern Africa*

Country	GDP/ Capita PPP$ 1993[1]	Human Develop- ment Index 1993[2]	GDP Rank Minus HDI Rank[3]	Calories/ Capita 1992	Rural– Urban Disparity 100= Parity[4]	Ginni Coefficient[5] 1=most unequal; 0=equal
Botswana	5,220	0.741	−10	2,288	74	0.54
South Africa	3,127	0.649	−6	2,705		0.55
Swaziland	2,940	0.586	−9	2,706		
Namibia	3,710	0.573	−37	2,120	39	0.70
Zimbabwe	2,100	0.534	−3	1,989	65	0.57
Lesotho	980	0.464	+21	2,201		
Zambia	1,110	0.411	+9	1,931	26	0.44
Malawi	710	0.321	+6	1,827		0.62
Angola	674	0.283	+1	1,840		
Mozambique	640	0.261	+2	1,680	94	
SSA	1,288	0.379		2,096	58	

Source: adapted from UNDP, 1996; EDC, 1997b
1 PPP$: purchasing power parity in US$ (allows for different costs of living).
2 Human development index: based on life expectancy, educational attainment and real GDP per capita (PPS$).
3 GDP rank minus HDI rank; a minus indicates the country has worse HDI indicators than would be expected from their GDP/capita (a lower than expected wellbeing given the richness of the country).
4 An average of disparity in health services, safe water and sanitation between rural and urban areas.
5 Ginni Coefficient is a measure of equality of income distribution.

(HDI). However, the richer countries in the region tend to have a lower HDI rank than would be expected from their GDP per capita (see Table 3.1). This is an indication that there is still considerable deprivation in these countries, despite their relative wealth, particularly in rural areas. There are high levels of inequalities in all countries in Southern Africa, although this is less pronounced in Zambia; the Ginni coefficients, which are a measure of inequality, are some of the highest in the world (see Table 3.1).

ROOTS OF RURAL POVERTY

There are at least three distinct approaches to the diagnosis of the twin problems of rural poverty and agricultural sustainability in the region (Wiggins, 1995):

- focusing on domestic political mismanagement, urban bias, etc;
- focusing on externally imposed constraints, arising from subordination to international capital in a globalized economy;
- focusing on the dynamics of agrarian change, as determined principally by the interaction of population growth and resource use.

Although it is simpler conceptually to look at these diagnoses separately, in reality rural poverty tends to be caused by a combination of factors.

Domestic Policy Mismanagement

This school of thought prioritises internal factors as causes of rural poverty, such as:

- price discrimination against agricultural produce;
- inefficient bureaucracies, insufficient or inappropriate investment in rural infrastructure, 'rent-seeking' tendencies in the state and local bureaucracy;
- stifling of rural entrepreneurial activity;
- weak rural institutions, resulting in poorly defined property rights and very high transaction costs.

Many reports have put particular blame on the political power of urban consumers, distorting policies against rural producers. Although price discrimination against farmers has been widespread in sub-Saharan Africa, the reality in Southern Africa is more complicated, and the interdependence of rural and urban households makes the diagnosis less clear cut:

- State intervention has gone beyond just keeping food prices low, and has included farm input subsidies, particularly for fertilizer and marketing subsidies through loss-making marketing boards (see Box 3.1).
- In some cases, support for certain types of farming, often involving the politically powerful farmers at the expense of the marginalised, has been as powerful a factor as the rural–urban divide. Thus in Botswana, influential (often urban-based) cattle owners have ensured a favourable policy environment for beef exports; and in Malawi, the estate sector successfully prevented the involvement of smallholders in profitable Burley tobacco growing for many years.
- Productive agriculture reduces poverty – not only to the extent that it supports rural livelihoods, but also by keeping food prices low, both for urban households and the many rural households who are net purchasers of food.

Political Culture and Governance

A development from the 'mismanagement' school of thought stresses other aspects of ineffective government, and has been the underpinning of more recent donor emphasis on programmes of good governance.

Within the region, there has been relatively weak accountability (particularly to rural constituencies) of both colonial and post-colonial governments, and active repression or cooption of autonomous development activity outside the control of the state. For instance, in Namibia, the cooption in many rural areas of traditional authority by the colonial government undermined its integrity and legitimacy. This process led to a severe breakdown of some community structures, and many communities degenerated from being adaptable, self-governing institutions into inflexible societies controlled by and dependent on centralised decision-making structures (EDC, 1997d).

Botswana probably has the strongest democratic tradition in the region, but even here government has been dominated by a relatively small cattle-owning establishment. South Africa has a strong and diverse civil society, but in rural areas the legacy of apartheid methods remains. Malawi acquired a reputation under President Banda for centralised but effective government, but government that was also coercive and stifled local initiative. Since 1994, the more plural and liberal environment in Malawi may be creating some new opportunities, but it has also resulted in a collapse of local discipline systems – manifested in the form of increased theft, deforestation, low credit repayment rates and an unwillingness to do community labour (EDC, 1997b). This indicates that the issue is not just removing state authority, but:

- conferring on the state greater legitimacy, by making it more responsive and accountable;
- (re)building local institutions (including attention to accountability, authority and capacity).

Paternalistic traditions in the region have been sustained by an unholy mixture of elite domination, the perceived imperatives of nation-building, fear of ethnic fractionalisation under a more liberal regime, and access to aid money motivated more by geopolitical than developmental considerations that insulated governments from having to be accountable to taxpayers. While these considerations have quite dramatically diminished, old habits and structures of government change more slowly. Popular democratic pressure for change in Zambia and Malawi, for example, has been quickly blunted by apathy borne of continued economic stagnation, including the difficulty of sustaining credible opposition without the powers of office.

Lack of Resources, Debt and Economic Structural Adjustment

Criticism of inappropriate domestic policies needs to be balanced by an appreciation of the difficult external conditions Southern African countries have inherited and have been forced to contend with. War, destabilisation, adverse changes to the terms of trade, drought and 'sunset clauses' (where the privileges gained during the previous regime were guaranteed for a period) have constrained development and reform in the various countries to different extents. Debt repayments absorb considerable proportions of revenue, and government service provision is being cut to reduce deficits as part of structural adjustment programmes (SAPs).

Analysing the effect of structural adjustment on poverty in the region is beyond the scope of this book. However, some points should be noted:

- Preadjustment policies produced distortions and inefficiencies within agricultural environments (see Box 3.2).
- Although adjustment has been needed in most countries in the region, the design and speed of change has often not been open to debate and local democratic accountability.
- External support, such as enhanced aid flows and debt cancellation, can ease the pain and make adjustment more likely to succeed.
- Some specific support is needed to help poor farmers to adjust to the new economic environment.

Structural adjustment programmes appear to have somewhat ambiguous effects on smallholder farming. SAPs have encouraged pricing that more accurately reflects true resource costs, thus avoiding static inefficiency and realising regional comparative advantages. The current liberalisation should in theory bring improvements to smallholder farmers. However, in reality it is not so simple:

- The adjustment is extremely painful for some farmers – poorer and remote smallholders are often least well placed to take advantage of the new opportunities, while still feeling the pains.
- Agricultural benefits for smallholders are sometimes swamped by other ramifications of structural adjustment, such as increased school fees, health charges, food costs and reduced off-farm income as household members are retrenched.
- Commercial (and sometimes NGO) alternatives to parastatal input and marketing channels may be slow to develop, may be monopolistic or may be concentrated in the most profitable areas.

The market-orientated environment does not necessarily create sustainable long term incentives. Additional emphasis on encouraging sustainability is likely to be needed (see Chapter 4).

BOX 3.1 – PRELIBERALISATION POLICY DISTORTIONS IN ZAMBIA

Before liberalisation in Zambia (and many other countries) there was a wide range of distortions which had different and often interacting impacts on agriculture.

(a) Overvalued exchange rates

The most important distortion: overvalued exchange rates roughly halved agricultural comparative advantage for those producers who could not circumvent them by exporting into neighbouring countries at black market rates (World Bank, 1992: 53).

(b) Food price controls

Domestic food crops were subject to additional distortions arising from direct price controls. Even at the official exchange rate, the import/export parity retail prices for maize during the 1980s were on average 40 per cent higher than the official price actually paid by Zambian consumers (World Bank, 1992: 12). Even if the exchange rate had not been controlled, domestic price controls still pitched the terms of trade heavily against farmers.

(c) Input subsidies

Subsidisation of fertilizer and credit encouraged farmers to utilise more of these inputs instead of unsubsidised land and labour. High levels of fertilizer use (but with low marginal returns) were consequently often combined with neglect of weeding and other labour operations. More contentiously, this may have fuelled the exceptionally high rates of urbanisation, or 'labour disinvestment at the farm level'.

(d) Pan-territorial pricing

The maintenance of a uniform maize price after 1974 had the effect of eliminating the comparative production disadvantage arising from differences in transport costs to and from markets.

(e) Commercialisation bias

A further distortion existed in favour of production for sale rather than for domestic consumption. The combination of pan-seasonal and pan-territorial pricing meant that many farmers in remote areas found it cheaper to sell their maize soon after the harvest and to purchase highly subsidised maize meal later in the year. The World Bank (1992:29) estimates that the proportion of total crop production sold onto the market grew from 54 per cent by value in 1980 to 77 per cent in 1988. This left farmers all the more vulnerable to subsequent changes in markets and prices.

(f) Lack of public investment

A final adverse incentive to improved agricultural performance arose from inadequate allocation of public resources to investment in infrastructure, principally in research, irrigation, electrification, and feeder roads (see Dumont and Mottin, 1979). These were doubly squeezed by a combination of low public-sector allocation, and the large share of that allocation going to subsidies and poorly performing parastatals. From 1981 to 1986, for example, agriculture received between 10.9 per cent and 15.1 per cent of the annual budget, but only between 4.1 per cent and 6.8 per cent per year were left over after payment of subsidies (Machina, 1996: 40).

Source: EDC, 1997c

Population Growth, Climate and Resource Degradation

There is a neo-Malthusian strand of analysis in each country in the region which points to rapidly rising populations (and also rising expectations of these populations) and contrasts this with the capacity of the natural environment to provide for these populations. Environmental degradation such as deforestation and soil erosion is seen as a consequence of this increasing pressure.

This approach has already been discussed in Chapter 1, with the conclusion that preventing a downward spiral of environmental degradation and poverty is possible but not inevitable. Sustainable intensification of agriculture, combined with the expansion of off-farm employment opportunities, can provide sustainable livelihoods given the right combination of technologies, community organisation and external environment; these are discussed further in the remaining chapters of this book.

CAN SUSTAINABLE AGRICULTURE PROVIDE A ROUTE OUT OF RURAL POVERTY?

With rural poverty endemic in the region, differing strategies are being proposed for eliminating this poverty:

- development of alternatives to agriculture (industry, mining, etc);
- commercialising smallholder agriculture;
- specific actions to address the needs of poorest smallholders.

It is not an either/or situation; a combination of approaches is required. Sometimes, however, there are hard choices to be made about where resources should be spent.

We need to move from thinking about agriculture and other means of livelihood as separate in order to understand the complementary role agriculture and off-farm income can play in household livelihood. Although agriculture alone is only likely to be a route out of poverty in some areas, it is still important everywhere. Treating on- and off-farm incomes as complementary will change the basis of our analysis – there is nothing wrong with agriculture producing 50 per cent of a household's livelihood as long as both off-farm and on-farm components are sustainable, complementary and can deliver an appropriate standard of living. Sustainable smallholder agriculture should therefore be a key component, although not the only component, of ending poverty in the different environments of Southern Africa.

This book argues that more priority should be given to specific policies and institutional support which address the needs of the poorest households. Given the diverse and complex nature of their livelihoods, these policies and services must be driven from the bottom up.

Commercialisation of smallholder agriculture is increasingly a policy objective of governments throughout the region: this is dangerous because it perpetuates the bias towards richer smallholders, and does not necessarily address sustainability issues (see Chapter 4). Commercialisation may be a viable prospect for those households with adequate resources; however, there is little evidence that it will work for less well-resourced households in less favourable environments. Commercialisation in practice usually means the increased use of bought inputs (fertilizer, hybrid seed, etc) and an increased concentration on cash sales, rather than production for home consumption. In policy terms, this encourages:

- agricultural extension services to provide more support to better-off smallholders (through marketing advice, etc);
- consideration within resettlement programmes to settle larger, better-resourced smallholders instead of those most in need of land (see Chapter 9);
- justification for policies more favourable to better-off farmers (eg giving exclusive grazing and water rights to better-off farmers and their syndicates in Botswana).

The commercialisation strategy often relies on the implicit assumption that agriculture alone needs to provide a complete household livelihood. In fact, as noted above, for many households, agriculture only provides part of the livelihood; this does not make agriculture any less important for these households – complemented by off-farm income, it can make the difference between abject poverty and an acceptable standard of living.

CONCLUSION

Rural poverty is acute throughout the region and is proving intractable, even in the relatively rich countries. The degree of inequality is also very high, with great differentiation even in individual rural communities. Rural poverty has been caused by domestic policy mismanagement, exacerbated by difficult external conditions including terms of trade, debt repayment and the legacy of war. Difficult climatic conditions are also a factor.

Agriculture is a vital component of livelihoods everywhere. More emphasis is needed on developing sustainable agriculture and off-farm income-generating opportunities in order to end poverty. The current emphasis in much of the region on the 'commercialisation' of smallholder farming, concentrating on more purchased inputs and marketing of products, is most suitable for better-resourced smallholders. However, it runs the risk of neglecting the needs of the poorer and more remote smallholder and of giving insufficient attention to the issue of sustainability.

Part II

Resource-Conserving Technologies

Chapter 4

The Damaging Bias Against Sustainability

INTRODUCTION

Technology is defined as 'the range of activities, including the physical means of production and farmers' own management practices, used by, or potentially available to, farmers to produce their desired output'. Technology can include 'traditional', 'indigenous' and 'modern' techniques – although in reality the division is not static. Farmers have always learned new techniques, either from their own experimentation or from their neighbours and, over time, these new techniques become traditional. It is only relatively recently that specialist research and extension services have contributed to this ongoing process of technology generation and dissemination.

Resource-conserving technologies are defined here as those that 'enable a farmer to produce her or his desired output, while maintaining the productive capacity for the future'. Choosing resource-conserving technologies may involve trade-offs between current yield and future yields, between current yield and the costs of producing it, and between current yield and risk. This book does not argue that a specific technological approach is best; instead, it contends that the technology choice should principally be left with farmers. However, it does argue that *the current technology generation and dissemination process, and the overall policy environment is, and has been for some time, biased in favour of short term production maximisation. This has encouraged unsustainable and high-risk smallholder farming, with detrimental consequences for poverty alleviation and the environment.*

Although the rhetoric may have changed in favour of on-farm research, farming systems, sustainability and smallholders, the reality is that most government research programmes in the region still concentrate on short term single-crop yield maximisation and increased use of external inputs. Often their recommendations are impractical for poorer smallholders. The reasons for this are quite logical:

- Research and extension on low external-input technologies are usually long term and more complex, and therefore tend to be neglected (see Box 4.1).
- Farmers need to feed their families today, and therefore often find it difficult to give sufficient attention to the long term.
- Most of the agricultural establishment has been trained within the yield-maximising, high external-input ethos.
- An increasing proportion of research and demonstration is being done by input suppliers, who naturally emphasise the use of inputs.
- Farmer organisations, who may advocate further research, often consist mainly of those smallholders with sufficient resources to use more inputs.
- Although NGO research is expanding and may be more orientated towards sustainability, it is still limited and often lacks a systematic and long term perspective.

Publicly funded research programmes need affirmative action in favour of sustainability to redress the current bias.

This book does not argue that the use of purchased or external inputs, such as inorganic fertilizer, is always wrong. Rather, it has been the focus on developing and promoting high external-input techniques that has led to the neglect of more sustainable techniques, which are often more appropriate for the poorest farmers. Southern African smallholder farmers – who usually know what is best for them – tend to consider research and extension's concentration on bought inputs as impractical and, since subsidies have been withdrawn, have consistently used less inputs than recommended, finding them too expensive and risky.

Paradoxically, due to the very low use of external inputs (such as inorganic fertilizer) in the region, there is generally scope for increased use of external inputs as well as a much greater emphasis on low external-input technologies. It is not an either/or situation – both are needed to achieve sustainable intensification (see Chapter 5).

Genetic engineering provides some controversial and powerful new tools which could result in damage or benefits to smallholders and consumers in Southern Africa. There is justifiable concern that:

- Genetically modified crops will be more suitable for richer farmers in more uniform environments, further undermining the competitiveness of smallholders in CDR areas.

BOX 4.1 THE PROBLEM OF THE LONG TERM

Sustainable techniques need, by definition, to work over the long term. For instance, an erosion-reducing tillage practice must be better than existing practice, when used over a ten-, 20- or even 50-year period (even if the practice is expected to evolve during this time). It may, however, have extra costs in terms of increased labour or reduced yield, and these often fall in the early years. Experiments or demonstrations over one to three years, which are typical in the region, are often irrelevant.

An example of the challenge of researching and promoting the long term is the leguminous tree *Faidherbia albida* (previously named *Acacia albida*), which may provide an appropriate way of maintaining soil fertility in some parts of the region. Its somewhat slow establishment means it may take 15 years to establish a decent stand, but this stand is then effective in maintaining soil fertility for 150 years (see Photo 4.1). How many research institutions, extension services or projects have this type of time perspective?

Farmers also need support to take the long term view. There may not be an incentive for farmers to adopt resource-conserving technologies until environmental damage seriously affects yields, and by then it may be too late since prevention is often easier and cheaper than the cure. Key issues for farmers are:

- Does the technology work over the long term?
- Are the returns in the long term adequate to compensate for the costs (land taken out of cultivation, labour, etc)?
- Will those investing reap the expected benefits (is tenure security adequate and what confidence can there be that the social and economic situation for farming in the future will give benefits)?
- How can farmers finance the investment in land improvement?

None of these questions can be answered with certainty; however, the creation of a truly enabling environment requires that we go considerably farther than at present to provide answers.

- Genetic material, preserved and developed over generations by small-holders, may be used without compensation by companies and then sold back to farmers in modified form at high cost.
- There may be unknown environmental and human health risks from genetically modified cultivars.
- Rules to regulate the biotechnology industry are being drawn up in the interest of Northern companies, farmers and consumers with little attention paid to the different needs and perspectives of Southern smallholders, consumers and companies.

While for commercial reasons it is likely that genetic engineering will exacerbate existing biases towards bought inputs (such as the use of specific herbicide and cultivar combinations), there are not necessarily biological reasons for this to be the case. Genetic manipulation *could* be done in favour of lower external inputs – for instance, by introducing the

Photograph 4.1: *Young* Faidherbia albida *tree growing in a maize field.*
These trees when large can help maintain soil fertility for 150 years, but careful
nurturing is needed in the first few years when growth is slow and the seedling
could easily be weeded out or stepped on.

ability of cereal crops to fix atmospheric nitrogen. Therefore, affirmative action is needed to:

- ensure that the interests and opinions of Southern African smallholders, consumers and companies are listened to when making the regulations.
- if appropriate within safety regulations, to ensure (probably with public sector funding) that the priorities of Southern African smallholders are included in future biotechnology research.

RESEARCH BIAS

Public Sector Research Bias

Public sector research continues to have a number of biases against sustainable smallholder agriculture (see Box 4.2). Although many institutions are genuinely trying to move towards a more appropriate research approach, they are constrained by:

- staff trained in narrow disciplines and specialised in on-station research;
- recruitment freezes, or lack of finance to hire new staff or retrain existing staff;
- budgets highly committed to existing establishment costs, with limited flexibility for developing new areas of work;
- 'short termism' – budget uncertainty (both from government and donors) and the three-year PhD cycle of researchers mean long term trials are difficult to establish and support; however, long term trials are essential for researching long term sustainability (see Box 4.1).
- lack of strong national rural development policy in favour of resource-poor smallholders and sustainability.

The end result is inappropriate recommendations. Farmers try to adapt recommendations to make them more realistic. Farmers often fail to follow research and extension advice, considering it too risky, too expensive or too labour intensive (see Table 4.1). It is interesting to note that personal communication with some senior staff of research institutions indicates that, although they defend their institution's recommendations, they often do not follow them on their own personal farms, modifying them in favour of lower inputs and reduced risk.

In most countries in the region, there have been genuine attempts to adapt research to smallholder needs. Much of this work has been done within farming systems units; while many of these units have produced some excellent work, they have usually found it difficult to enter the mainstream and to transform either the overall research agenda or the

---

BOX 4.2 KEY PUBLIC SECTOR RESEARCH BIASES

(1) The majority of experiments are run for a short time period (one to five years) and are designed to provide short term recommendations. While most institutions acknowledge the importance of sustainability in their reports and plans, experiments looking at sustainability are additional, and often peripheral, to the work of the research institution. Practically no institutions use an approach in which long term sustainability is a factor in all relevant experiments.

(2) The majority of research still has the objective of production or yield maximisation with little attention paid to other trade-offs. Relatively few experiments are designed to find either financial or economic optimum combinations of inputs and yield. Even in land surplus areas, nearly all crop experiments are designed to reveal yield per hectare, rather than yield per unit of labour (which is often the more relevant constraint).

(3) Very little attention is paid to risk minimisation or the balance of achieving high production with acceptable risk. In reality technologies need to produce sustainable livelihoods that can weather severe set-backs.

(4) Similarly, despite over a decade of nominal adherence to a farming systems approach, most institutions remain organised along commodity or discipline lines. Many institutions have a farming systems unit as an add-on rather than a guiding approach to all their work (this is about to change in Zimbabwe). Crops, livestock, forestry and wildlife are often responsibilities of different institutions, with limited collaboration. Farming system work often tends to be donor driven and funded, rather than part of the core budget and staff of the institution (see Box 4.3).

(5) Agricultural economists and rural sociologists are underrepresented (or non-existent) in most institutions. Where they do exist, they are often not used strategically and may be marginalised.

(6) Many agricultural research stations are situated on favourable soils or in higher rainfall areas and are therefore not typical of smallholder conditions.

(7) There is still relatively little consideration of gender in most research programmes and, where included, gender is often not integrated into the overall approach.

(8) In most institutions there has been a shift in policy towards smallholder participation, but, in practice, it is mainly the better-resourced, larger-scale smallholders that are involved in trials, field days and represented on committees.

(9) With budget cuts, on-farm research tends to be hardest hit (often because of pressure on transport budgets).

---

national policy framework. Bias has continued. An example is the Adaptive Research Planning Team (ARPT) programme in Zambia (see Box 4.3). Another example is the ALDEP programme in Botswana (see Box 11.12), which had a budget of 22 million Pula between 1985 and 1991. This was swamped by the much less sustainable Accelerated Rainfed Agricultural Programme (ARAP), which spent 190 million Pula in the same period.

Table 4.1: *Comparison of Cropping Recommendations and Farmer Practices in the Semi-Arid Areas of Chivi, Gutu and Sanyati of Zimbabwe*

Management	Agritex Recommendation Practice	Farmer Practices
Crop varieties	Specific varieties recommended, usually hybrids or other improved varieties. Maize in NR III and IV, small grain in IV and V.	Recommendations followed for maize and cotton, with local varieties for sorghum and millet. Maize widely grown in all natural regions.
Land preparation	Winter plough (June–July) to 20–25 cm. Apply manure (August–September).	5% of area winter ploughed to 10–15 cm. 6% of area manured.
Planting dates	Planting with first effective rains (November and December).	Staggered from dry planting of millet in October to late planting of replacement crops in February and March. Dry planting vleis with groundnuts or maize and double cropping where conditions allow.
Intercropping, relay cropping, late planting and in-filling	No recommendation.	Up to 75% of maize intercropped. Relays following crop failure important. In-filling and late planting widely practised.
Fertilizer applications	*Grain crops* 100–200 kg D/ha 50–100 kg AN/ha *Cotton* 150–250 kg L/ha 50–100 kg AN/ha Eight tonnes manure per hectare. 1 bag per nine metre intervals and shallow ploughed.	>10% of area with basal fertilizer often after planting. >1% top dressing with AN. Low concentration, spot applications for certain crops (maize and cotton), otherwise none. 6% of area with manure applied. Low application rates, careful rotation, focused on certain soils.
Weeding and cultivation	Two to three weedings in first eight weeks.	*Maize* 20% has no weeding. 60% of area has one weeding. 20% weeded twice. 1% weeded three to four times. *Cotton* 80% weeded two to three times.
Pest and weed control	*Maize* Diptrex and Thiodan *Cotton* Various	No pesticides except on cotton when up to nine applications can be used. Some indigenous remedies may be used.

Rotations	Maize with manure, maize/ sorghum, legume, millet.	Mostly continuous maize. In Sanyati, maize continuous cotton.
Soil and water conservation	Tied ridges No-till tied ridges Ripping into crop residues, tine planting or direct drilling Potholes	Less than 1% adoption. Ridging in vleis or heavy soils. Potholes in high fertility areas such as anthills.
	Contour bunds, waterways and storm drains, contour planting, green manures, cover crops.	Contour bunds, contour planting. Micro-scale water harvesting using contours or directing water to plants

Information from:
(a) Agritex budgets (1993 and 1994) and personal communication with extension workers
(b) Chivi 1990–91; FSRU (1994); Farmer monitoring Munyaradzi (Gutu) and Nyimo (Sanyati) 1992–93 and 1993–94; and personal communication with farmers
Source: Ellis-Jones and Mudhara, 1995

Commercial Sector Research

An expanding proportion of overall research is undertaken by the commercial sector, particularly by seed and agrochemical companies. Data are scarce, but Beynon (1996) estimates that 15 to 50 per cent of agricultural research in Zimbabwe is by the private sector. This research can be useful and of high quality, but it has very specific orientation:

• in favour of bought external inputs and packages;
• in favour of farmers of size and type able to buy inputs at an acceptable transaction cost.

Some of the research funded by the commercial sector is actually done by the public sector; for instance, South Africa's Agricultural Research Council is now expected to raise a proportion of its overall budget through commercial contracts. This contracted research tends to have the same orientation as other commercial research, despite being carried out by the public sector.

Public sector research needs to take into account the work done by the commercial sector, with the objective of redressing the imbalance. This is a reason for the public sector to give additional priority to low bought-input techniques and to the type of smallholders not served by the commercial sector. Some research in the region is funded by levies from certain crops (tea, tobacco and cotton). While, in theory, this research should be demand driven, with farmers involved in prioritising research areas, it is usually the better-resourced and often large-scale farmers who are active on these boards and, in reality, research is often directed towards their needs.

Box 4.3 ARPT, Zambia

There is a long tradition in Zambia of reorienting agricultural research from a market- to a livelihood-centred perspective, notably through attempts to institutionalise farming systems research (Kean, 1993). At the vanguard of this process was the Adaptive Research Planning Team (ARPT), set up within the research branch of the Ministry of Agriculture, Fisheries and Food (MAFF) in 1980 (Kean, Sutherland, Sikana & Singogo in Wood et al, 1990). ARPT grew steadily during the 1980s, setting up provincial units, seeking to influence agricultural research priorities to reflect smallholder priorities, and encouraging other agencies to plan their work around a more holistic understanding of farmers' livelihoods.

However, poor financial returns, and the political weakness of those who stood to benefit from the research, have hindered the process, particularly in the context of public sector stringency and aid cuts. ARPT links with other research teams in MAFF were also weakened by suspicion and fear that it was diverting resources from more orthodox agricultural research. ARPT was overreliant on donor funding and failed to establish its impact, let alone social cost-effectiveness, resulting in weak farmer demand–pull. The result has been that most public sector agricultural research has reverted to being directed primarily towards the demands of richer and more market-oriented farmers.

Meanwhile, ARPT sought to strengthen farmers' capacity to carry out their own research through promoting village research groups, with the ultimate goal of making these initiatives 'part of a self-generating process resulting in demand-led agrarian change' (Sikana, 1994: 244). While laudable in intent, this approach raises unanswered questions about the conditions (including duration, form and affordability of external support) for ensuring that such groups become durable and effective, rather than being founded on unrealistic expectations (Dobel, 1996: 22). Part of the answer perhaps emerges from analysis (EDC, 1997c), suggesting that small farmer groups built around a single activity are unlikely to be as sustainable as a hierarchy of groups flexible enough to accommodate a range of different and sustained activities, including links with external agencies.

Source: EDC, 1997c

Non-Governmental Organisations and Research

NGOs are increasingly involved in research in the region, although the distinctions between trials, demonstration and promotion in some NGO work can be blurred. Sometimes NGO research has a specific focus, such as ZIP Research (linked to the Zimbabwe Institute of Permaculture), Veld Products Research in Botswana (see Box 5.4) or ENDA's work on small seeds in Zimbabwe (see Box 11.5). In Mozambique, World Vision International has a large research programme, covering much of what is usually a government research responsibility.

The quality of NGO research in the region is variable, ranging from excellent to irresponsible. Some NGOs are narrowly trying to prove and promote a particular technique, rather than widen choices for smallholders, while others have a more open agenda. Occasionally NGO research is being done in collaboration with the public sector – which can result in useful synergy between the NGO's orientation towards poorer smallholders and the public research institute's technical competence. This can also have the useful effect of bringing the results to the attention of both government and NGO networks.

There are, however, a disturbing number of cases of poor communication between NGO and public-sector technology development, and some examples of open hostility by public sector institutions to alternative technology that is promoted by NGOs. An example of the latter occurred in Malawi, where the promotion of A-frames, Magoye soya beans and vetiver grass were all originally opposed by government institutions and had to be promoted through NGO networks, sometimes in the face of open hostility from government research and extension. Fortunately, since some of the advantages of these technologies have become apparent, government policy has changed and they have now been adopted and promoted by government.

In Zimbabwe, some NGOs maintain a distance from government research and extension, believing they are insufficiently orientated towards equity and sustainability issues. However, recently, people involved in government agricultural research and extension are showing interest in alternative approaches, for instance by attending courses in permaculture – therefore, relationships seem to be thawing.

Where relations between government and NGOs over technological choices are poor, part of the problem often lies in the perceived role of research and extension. If the institution sees its role as producing recommendations for farmers to follow then alternatives to these are seen to be threatening. If, on the other hand, the institution sees its role as producing options for farmers to choose from or supporting farmers own experimentation then it may be more open to the initiatives of others.

EFFECTS OF TECHNOLOGICAL BIAS

Bias in favour of agricultural yield maximisation has had a number of consequences:

*   High-input techniques, particularly the use of inorganic fertilizer and hybrid seed, have been researched and promoted at the expense of low external-input alternatives.
*   Inputs have often been subsidised through grants, subsidised prices, subsidised credit and subsidised sales channels.

Subsidised inputs have encouraged farmers to use more bought inputs than they otherwise would have done. This has, however, caused a predictable crisis in some countries, when liberalisation has resulted in the sudden withdrawal of subsidies to which farmers had grown accustomed. This has made inorganic fertilizer uneconomic on cereals – and yet alter-

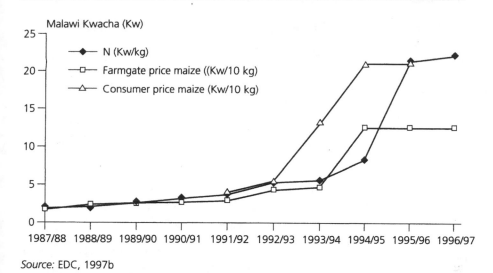

Source: EDC, 1997b

Figure 4.1: *Relative Price of Fertilizer and Maize in Malawi*

natives are underdeveloped. The problem is particularly acutely in southern Malawi, where land shortage is causing almost continuous cropping with maize. In Malawi, the fertilizer price has risen twice as much as the farmgate price of maize; therefore, while in 1992 it took ten kilogrammes of maize to pay for one kilogramme of nitrogen (N), in 1996 it took 22 kilogrammes. Since the typical response to N under smallholder conditions is 15 to 20 kilogrammes extra maize per kilogramme N used, it is no longer profitable to use fertilizer on maize (see Figure 4.1).

In Malawi, the change in fertilizer profitability caused a fall in the previously upward trend in fertilizer use, but it was initially masked in 1991/92

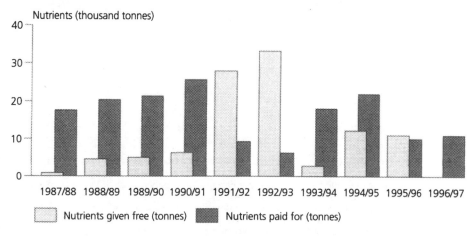

Sources: Conroy, 1993; Government of Malawi, 1996; Malawi Rural Finance Company

Figure 4.2: *Nutrients Used by Smallholders on Maize in Malawi*

Table 4.2: *Smallholder Purchases of Fertilizer and Seed in Zambia*

Season	Number of Smallholders Purchasing Fertilizer	Tonnes of Fertilizer Purchased	Tonnes of Seed Purchased
1990/91	366,000	122,000	17,000
1993/94	166,000	72,000	5,500

Source: GRZ/CSO, 1991; 1994

BOX 4.4 THE ECONOMICS OF USING FERTILIZER ON PEARL MILLET IN NORTHERN NAMIBIA[1]

| | Namibian $ (1996/97 prices) | | |
	No Fertilizer	Subsidised Fertilizer	Commercial Fertilizer
Value of pearl millet (threshed) – 300 kg/ha (no fertilizer) and 750 kg/ha (fertilizer) @ N$ 1.60/kg	**480**	**1200**	**1200**
Seed (Okashana-subsidised)	15	15	15
Fertilizer –			
3 x 50 kg bags basal 2:3:2 (30%)	–	64.5	306
2 x 50 kg bags urea topdressing	–	32	189
AgriBank Interest @ 13%[2]	–	15	66
Incremental labour – 28 days @ N$ 10[3]	–	280	280
Total costs[4]	**15**	**386.5**	**856**
Net Value	**465**	**813.5**	**344**
Variable Cost Ratio		**1.86**	**0.84**

Source: EDC, 1997d
1 Figures used in this box derive from production figures and prices in the north central regions of the Northern Communal Area, assuming good rains and that draught-animal powered tilling and weeding is used, as presented in a draft paper prepared by the RDSP (RDSP, 1997), and from retail fertilizer prices from Agra Grootfontein.
2 The National Agricultural Credit Programme's interest rates are due to rise annually until parity with commercial interest rates is achieved.
3 Assuming a combination of family and hired labour.
4 This only includes the incremental costs of labour, draught, etc due to the use of fertilizer.

and 1992/93 by free distributions as part of drought relief, and by farmers reneging on payments as the credit system collapsed. However, the trend in the amount farmers actually purchased is striking (see Figure 4.2).

Similar trends have been experienced in other countries in the region – for example, Zambia (see Table 4.2).

Fertilizer remains subsidised for communal farmers in Namibia, (although a survey in 1992/93 showed only 3 per cent of farmers in Northern Namibia using it). The objective is to encourage usage among better-resourced smallholders, presumably before removing the subsidy (since there is a more general policy of phasing out subsidies). A sample field budget shows that this policy is difficult to justify; without the subsidy there seems little reason to use fertilizer: it is more profitable in an average year to grow millet without it (see Box 4.4). The disadvantage of using fertilizer would be even greater in a drought year, with the farmer risking losing all his or her expenditure.

With current prices, inorganic fertilizer is economic only in some areas and for some crops. Extension services throughout the region have been slow to internalise the changed economics of fertilizer and have continued to promote it, even when it is proving to be uneconomic.

Demand for Resource-Conserving Technologies

In theory, a way of ensuring that research priorities are more appropriate to smallholders is to have strong and influential demand from smallholders. However, effective, articulate demand for appropriate research from the poorest farmers is often weak.

It is important to be realistic and to recognise that farmer demand does not necessarily prioritise long term sustainability – smallholders, like many others, are often forced by circumstances to prioritise short term constraints. For many farmers, resource-conserving technologies may not become a priority until deterioration of the environment (eg erosion) affects yields. Enlightened governments, donors and NGOs have a continuing role in ensuring that research has an orientation towards long term sustainability and an environmental policy is created which enables farmers to make longer term investments in sustainability.

More detailed research is needed on the uptake (or lack of uptake) of potentially beneficial technologies. In most areas there are:

- examples of seemingly beneficial technologies with very low uptake;
- examples of individual households using a range of innovative technologies, which are often not being adopted by their neighbours; the technologies, coupled with extremely hard work, skill and enthusiasm, result in these households producing considerably more than their neighbours – some of these 'islands of excellence' are internationally famous.

Labelling individuals who use these technologies as progressives and, therefore, by implication the rest as laggards, can miss the point that there are often very real constraints to adopting some technologies by households (very often labour shortages at key times). Understanding the reason for low adoption can lead to changes either in the technology or in the external environment (such as the need for a grant scheme to support farm level investment – see Box 11.12).

CONCLUSION AND RECOMMENDATIONS

Bias in favour of purchased inputs and the short term has led to underdeveloping and underpromoting potentially more sustainable alternatives. Public sector research (government and NGO) needs to complement research in the commercial sector in order to redress the balance with affirmative action in favour of poorer smallholders and sustainability.

BOX 4.5 AFFIRMATIVE ACTION: TECHNOLOGY DEVELOPMENT IN FAVOUR OF SMALLHOLDER SUSTAINABILITY

(1) Clear priorities in favour of smallholders and sustainability.
(2) Long term perspective to make possible experiments into the maintenance and enhancement of yield over the longer term (implies a degree of funding security).
(3) Active creation of a learning environment involving farmers, extension and research, with research and extension stimulating, enabling and publicising research done by farmers.
(4) Providing options for farmers to choose from rather than recommendations.
(5) Appropriate reward system to reward researchers for adopting technologies produced, rather than for yields obtained on station, or papers published, etc.
(6) Research planned in accordance with the priorities of smallholder farmers, including women, and with an understanding of the overall farming system.
(7) Adaptive locally based research, responsive to diverse environments.
(8) Research not only looking at how to increase yield, but how to reduce costs, reduce labour, reduce risk and reduce environmental damage.
(9) Multidisciplinary approach, making use of multidisciplinary teams, with each member having a holistic perspective in addition to their specialisation.

There is a need to address both the potential benefits and the potential risks from biotechnology, with affirmative action to:

• Ensure that the interests and opinions of Southern African smallholders, consumers and companies are considered when making the regulations.
• Ensure, where safe, that the priorities of Southern African smallholders are included in future biotechnology research (probably with public sector funding).

Chapter 5

Some Resource-Conserving Technologies

INTRODUCTION

This chapter discusses a number of technologies that may be useful in different circumstances to enable smallholders to increase production on a more sustainable basis. However, it is important to remember that technology is context dependent; what works in one place, or under a particular set of circumstances, will not necessarily work elsewhere. The technologies are mentioned as possibilities. Ultimately, farmers need to be provided with a basket of choices from which to choose.

The technologies are discussed for convenience under a number of headings; however, at the farm level, different technologies interact. It is often the way technologies combine or conflict that makes the difference between success and failure.

TECHNOLOGIES FOR SUSTAINABLE PRODUCTION

Optimising Water Use, Soil Conservation and Fertility Management

In drier areas, soil moisture is often the limiting factor, so that fertility improvement is only effective when measures are also taken to reduce moisture stress. In the past, much of the extension emphasis has been on the safe removal of water from fields to prevent erosion, rather than its more efficient use for crop growth. Clearly, there is a need for techniques for combining both.

There is a range of techniques for improving water-use efficiency – tied ridges, tied furrows, micro catchments, infiltration pits, etc. The relative benefits of these tend to be quite site-specific, depending on slope and soil texture. For instance, experiments over a number of years suggest that in the driest regions of Zimbabwe, tied furrows increase sorghum yields by 4 to 79 per cent in clay soils, 29 to 100 per cent in medium-textured soils and have no or even a potentially deleterious effect in sandy soils. This implies a need for flexible extension, offering farmers alternatives and empowering farmers to experiment and choose. Where tied ridging is appropriate, it is often best done as a combined operation with first weeding to limit the additional labour required. Soil variability can be exploited, with the selection of micro environments with heavier soils, or where water collects, for particular crops, even within a field; this can increase overall yields significantly.

Reduced tillage residue and mulch farming, no-till ridging and no-till strip cropping have all been shown to reduce soil loss in Zimbabwe (Elwell, 1996). In practice, there are some difficulties for farmers in adopting these:

- It is often difficult for smallholder farmers to maintain the 30 per cent mulch cover necessary to protect the soil from erosion, since yields are low and livestock browse the crop residues in the dry season.
- Weed competition and increased labour needed for weeding can outweigh benefits.
- Some of the benefits from reduced erosion may be felt communally (eg reduced siltation of dams), but the costs fall on the individual.
- Germination is sometimes poor in reduced till systems.

Low and often declining soil fertility is (after moisture stress in dry areas) the major constraint to increasing crop yields. Soil fertility declines under cultivation because of:

- nutrients removed in the harvest;
- erosion of fertile topsoil;
- leaching of nutrients below rooting depth; and
- mineralisation of organic matter when exposed to the sun and air on cultivation.

Traditionally, fertility was maintained through fallowing; with increasing land pressure this has tended to be replaced by almost continuous cropping, sometimes combined with the use of bought inorganic fertilizer, termitaria soil, leaves from nearby forests, *kraal* manure, and crop residues. Compost and household residues also replace some fertility. Where inorganic fertilizer is not available or too expensive, it is difficult for techniques using local materials to be effective on a whole field basis because of labour constraints and the shortage of suitable material; therefore, these techniques are often concentrated on vegetable plots or fields near the homestead, with a conse-

Photograph 5.1: *Spreading* kraal *manure on fields can help fertility – but in order to have a significant impact, considerable quantities are needed, which may not be available and requires transport and labour.*

quent fall in fertility in the main fields (see Photo 5.1). Farmers are also sometimes wary of using manure because of the dangers of crop burn in dry conditions and the introduction of weeds.

Farmers often use an opportunistic approach to fertility, making optimal use of local sources first before resorting to expensive bought fertilizer. In addition, rather than use a basal dressing, the farmer may wait and see how good the rains are and then top dress as necessary. This approach is logical to reduce risk – and research needs to be designed to optimise this type of strategy.

Some proponents of sustainable agriculture advocate organic farming (avoiding all inorganic fertilizer and other manufactured chemicals) as the only truly sustainable way forward. This can be appropriate where farmers have sufficient organic material to replace nutrients; however, this is not *currently* the case for most smallholders. Intensive cropping, with insufficient nutrient replacement, leads to less vigorous plant growth, resulting in less organic matter being returned to the soil (either the end of season decay of vigorous root growth and plant remains, or the dung of livestock eating these, remains) and less ground cover to protect the soil from rain and wind erosion. Therefore, *incorrectly pursued*, organic farming can lead to more environmental damage than moderate use of inorganic fertilizer. For this reason this book advocates an appropriate mix of means of maintaining soil fertility – with what is appropriate being decided by local farmers according to local circumstances.

Agroforestry, the combining of trees with crops, can provide a certain degree of fertility maintenance, along with other benefits derived from

Photograph 5.2: *Multipurpose trees growing in crop land can bring many advantages – but it can be a struggle to protect them from grazing livestock which come into the field after harvest – here a stockade of poles is being used, indicating considerable tree cutting to protect one planted tree.*

trees. However, there are some trade-offs. The inclusion of enough trees to have a major impact on soil fertility (eg in alley cropping) is likely to reduce space for crops and have labour implications. Less intensive inclusion of trees on contour ridges and field boundaries may be less intrusive, but the benefits will be lower. In addition, trees that produce other benefits, such as fruit or poles, may be less useful in fertility maintenance than single-purpose trees. The problem of fields being used for dry season grazing, and consequently trees being destroyed by livestock, is a limitation in many areas (see Photo 5.2). There are no magic bullets, but there are opportunities for offering a range of choices to farmers.

Table 5.1: *Technologies for Soil, Water and Fertility Conservation*

Objective	Techniques	Comments
Soil and water conservation	**Infield water management (physical):** contour ridges and furrows (tied, untied), contour cultivation, contour strip cultivation, aligned by A-frame or other appropriate level (see Photo 5.3). Infiltration pits and micro catchments.	Can reduce soil, water and nutrient loss significantly if correctly done. Uptake depends on how easily it fits into existing farming system and how much labour involved. Multipurpose trees can be planted on ridges. Some examples of traditional precolonial terracing (eg Inyanga terraces in Zimbabwe). Some examples of farmer opposition due to enforced terracing, particularly during the colonial period.
	Infield water management (biological): ways of ensuring that the maximum benefit is obtained from the limited water available.	A variety of micro-management techniques are used or have potential. Examples include within field variations in cropping density and crop type, timely and staggered planting, effective weeding and intercropping. Short season varieties and crops. Early and regular weeding is particularly important in dry areas and techniques to reduce labour requirement can be effective.
	Vetiver grass strips and other **barriers across fields** and in gullies. *Pennisetum typhoides* may be more suitable grass in drier areas.	Becoming part of mainstream practice. In some areas there is a shortage of planting materials. Adoption can require community mobilisation. Adoption often depends on perceived benefit to effort ratio. Rehabilitating gullies may be dealing with the symptom rather than the cause of the problem
	Reduced tillage with crop residue left on surface as protective mulch.	Lack of soil disturbance maintains soil structure and reduces oxidisation of organic matter, relies on adequate residues to protect soil surface from rain. Low yields mean many smallholders have insufficient mulch to be effective, or residues are eaten by livestock, or used for alternative purposes.
	Water conserving irrigation techniques.	Techniques such as upturned bottles, clay pipes, trenches, trickle irrigation, etc often linked to effective mulches have potential, especially on the small scale.
	Water development: dams, vleis, haffirs, catchment tanks, wells, sand river extraction.	The construction of more but smaller dams and their local management for irrigation or livestock is one area of potential throughout the region. There is potential for more sustainable use of naturally occurring wet areas (vleis), however sometimes legislation is out of step with local practice and opportunities (eg Zimbabwe – EDC, 1998a). Appropriate affordable water extraction and

storage techniques require more development and information exchange about lessons learned.

Soil fertility maintenance	Increased use of **inorganic fertilizer,** usually combined with fertilizer responsive or hybrid varieties.	This has been the main thrust of crop research and extension. Despite this, adoption rates remain generally low. Unfavourable grain, fertilizer prices, particularly since economic liberalisation, and risks due to drought have reduced the sustainability of this approach.
	Intercropping with **nutrient enhancing trees** such as *Faidherbia albida*.	*F. albida* is promising in many areas with possible long term (150 years) contribution to soil fertility and low management/labour. But slow establishment and time to start giving results (five to 15 years) discourage farmers and researchers. Direct seeding may improve rapid establishment.
	Alley cropping and mixed tree intercropping with coppicing of species such as *Leucaena,Sesbania, Gliricidia, Tephrosia, Sena*.	More promising in moister areas. Initial high expectations from research stations modified with realisation that considerable management and labour input are needed to avoid shading and depressing of crop yields. Careful species choice needed according to site. Also can produce other benefits such as poles, firewood, fodder.
	Relay cropping/ under-sowing with leguminous species such as *Tephrosia* at first or second weeding.	Fertility-enhancing crop grown as 'follow-on' after cereal crop is harvested. More viable in less dry areas. Promising early results in Malawi.
	Improved fallow by planting fast-growing N fixing tree or bush, such as pidgeon pea or *Sesbania sesbans*, leaving for one to three years, followed by normal cropping.	Good research results from Zambia with yields increased for several years and positive overall returns. Way of restoring degraded land and can provide poles, firewood, fodder, etc. Lower production in first years due to land taken out of cropping makes adoption difficult for land hungry households.
	Legume rotations and inter-cropping: soybean, pidgeon pea, cowpea, *phaseolus*, groundnuts, bambaranuts, etc.	Producing crop of increased food value and maintaining soil fertility. Self-inoculating Magoye soybean popular introduction in Malawi, producing high yields of quality food. Cooking demonstrations helpful in improving digestibility and acceptability (see Photo 5.4).
	Use of ***kraal* manure and compost.**	Although used throughout the region, there is potential for increased use as soil fertility becomes more critical. Direct incorporation of crop residues into soil can immobilise nitrogen. Often transport of manure from the *kraal* to fields is the limiting factor; carts can help. Crop residues introduced into the *kraal*, and later used on fields, can provide additional feed, absorb urine and reduce nitrogen immobilisation.

Photograph 5.3: *An A-frame level – in many areas agricultural extension does not have sufficient levels of staff to peg contours – the A-frame provides an appropriate alternative,which can be made and used by the farmers themselves.*

Weeds are a major constraint to improved yields (especially lack of early weeding) since weeds compete with the crop for scarce water and fertility. Lack of labour for weeding is a constraint to increasing the area planted and to introducing reduced tillage. A number of techniques, some of which can be used in combination, can reduce weed impact:

Photograph 5.4: *Women threshing soya beans in Malawi – this variety, Magoye, is particularly suitable to smallholder cultivation, as it does not need to be innoculated artificially with rhizobium in order to fix nitrogen.*

- choice of varieties – some are more sensitive to weed competition than others;
- rotations and intercropping (eg cowpea can reduce *Striga* impact in cereals);
- improved implements (eg weeders with reduced draught require-ments suitable for donkeys);
- granular atrazine weedkiller (in conjunction with minimum tillage) – this might be seen as environmentally hazardous; however, the benefits of reduced erosion from minimum tillage would need to weighed up against the cost of the chemical and any damage caused.

Crop Diversification and Varieties

Crop diversity can reduce risk, yet there are indications that the diversity of crop species and varieties grown by smallholders has been diminishing in recent years. There is ongoing debate about the merits of maize, compared with small grains, and there is concern about a continuing bias in favour of maize at the expense of more drought-resistant small grains (principally sorghum and millet):

Table 5.2: *Sustainable Crops, Varieties and Value Added*

Objective	Techniques	Comments
Increased yields and risk minimisation	Growing more **drought-resistant crops,** such as small grains and short season varieties. Growing more cassava and sweet potato.	Efforts are being made to counteract perceived bias against small grains (see Box 11.5). Major breeding successes with short season varieties.Cassava spreading in some areas; promotion is controversial because of low nutrition content and its heavy depletion of soil fertility. Nutrition can be improved if leaves are eaten as well as the roots.
	Diversity (1): planting on range of planting dates, using range of varieties and species, in range of scattered fields. Intercropping.	Reduces impact of drought and pest attack. All these are traditional techniques that have tended to be ignored by extension, and some are declining in use.
	Diversity (2): smallholders increasing production of existing cash crops – sunflower, cotton, tobacco, sugar.	These crops are not necessarily resource conserving as they may be erosion prone, heavy users of nutrients or agrochemicals. However, they can, if grown appropriately, make livelihoods more sustainable and make investment in the farm possible. Vegetables and fruits provide a small, but growing, contribution to smallholder livelihoods throughout the region. Smallholders need to adapt to different market conditions with caution against overreliance on a single crop or market. There may be a gender dimension – with men using land and resources for cash crops at the expense of food crops – which may be considered the women's responsibility.
	Diversity (3): promotion of new alternative cash crops – chilli peppers, bixa (colourant), *Trichilia* (biodiesel), jatropha, castor bean, etc.	Some may have significant, even if limited, potential. There is unlikely to be a single magic bullet.
	Development of **natural pesticides, biological control,** integrated pest management.	Research into both indigenous technical knowledge (ITK) and into alternative natural pesticides is taking place throughout the region, but has rarely entered the mainstream. Insecticidal properties of tobacco, *Neem* and *Tephrosia* are being used on a limited scale. Ash can be an effective alternative to Dipterex in control of stalkborer. Biological control agents have so far been little used for smallholder crops; an exception has been a wasp to control the cassava mealy bug.

Local Value Added	On-farm or community **processing.**	Particularly important for those smallholders excluded from liberalised markets. Some examples are a robust oilseed press for sunflower made in Zambia, a mill for small grains made in Botswana, the production and local selling of sour milk, and the local processing of skins.
	Household or community level **grain storage** or cereal banks. **Seed banks.**	Continued work is needed on appropriate on-farm processing and storage. Cereal banks have potential for reducing differential between buying and selling prices (important for the many households who both sell and buy). Many community-level initiatives have had low sustainability due to management difficulties and decapitalisation in drought years. Seed banks sometimes provide a structure through which to introduce improved varieties and to conserve and spread existing varieties.
	Improved **blacksmith technology.**	Can produce locally available, better tools and repairs at cheaper price. ITDG have a programme supporting blacksmiths in Zimbabwe.

Table 5.3: *Techniques for Reducing Labour Demand*

Objective	Techniques	Comments
Reduced labour	**Tractor power**	Institutional schemes run by government or community/NGO have not generally proved cost effective or sustainable. Private hire arrangements are spreading throughout the region but can be expensive or untimely for poorer smallholders.
	Animal traction	Both for ploughing and weeding. The latter requires row planting, which can increase labour demand at a crucial time. Improved weeding using a cultivator has had limited uptake in some areas. Use of animals for transport of crops to market and manure to the fields is important in parts of the region. Poorer households have declining access to animal draught since many lost livestock in recent droughts. There have been limited restocking attempts.
	Improved stoves (reducing firewood use and therefore wood collection time)	Despite good theoretical advantages and many promotion programmes, there has been limited spontaneous adoption, partly because of values of fire beyond cooking – for light and to sit around socially.

- research has concentrated on maize.
- marketing has concentrated on maize.
- extension and input supplies have concentrated on maize.
- local grinding mills have been suitable only for maize.
- drought relief food and seed distribution have concentrated on maize.

Many households have grown accustomed to maize as the staple diet and fewer small grains are being grown by farmers. Drought has led to the loss or decline of local small-grain seed landraces.

There have been a number of initiatives to try and counteract bias against small grains, with only moderate success (see Box 11.5). Farmers' interest in growing maize is not irrational and not only due to bias within the institutional environment:

- Small grains produce less yield than maize in a good rainfall year (although in a drought year they can produce some yield when maize fails).
- Small grains are vulnerable to bird damage (particularly with children going to school rather than bird scaring).
- Small grains can require more labour to process and can be difficult to sell and unacceptable at local grinding mills.

However, the droughts of the early 1990s showed the danger of overreliance on maize – particularly in the more marginal drier regions. In these drought-prone areas, diversity is the key strategy; ideally, small farms in these areas should be planted with a mixture of crops and varieties, including both maize and small grains – as well as other crops – in order to obtain a yield in bad years, as well as good. Interestingly, recent interviews in southern Angola indicated that farmers were consciously reducing their dependence on maize and returning to small grains because of recurring droughts in the area (Pantuliano & Whiteside, 1997).

Much breeding and extension has concentrated on hybrid maize; however, there are reasons to continue work on open-pollinated varieties (OPVs). Improved OPVs of sorghum have been developed recently which can increase yields by more than 50 per cent (SADC/ICRISAT, 1993). However, these varieties, which are early maturing, are sensitive to fungal attack if the weather is humid during maturation; thus, what is good in a drought may be worse in a good year – confirming the need to grow a mixture of varieties. Analysis also shows that some traditional varieties have considerable potential and need to be conserved and made more widely available. Interest has been shown by farmers in the Chivi district of Zimbabwe in open-pollinated varieties of maize. Under difficult growing conditions yields of improved OPVs and hybrids are similar, but the OPVs have other advantages, such as cheaper seed and better storage. A problem is that there is often little incentive for commercial seed companies to develop and sell improved OPVs: after the original purchase farmers can continue to retain seed and share it with their neighbours, cutting company profits. This indicates a continual need for public or NGO sector activities in this area.

Most research and extension continues to focus on monocultures; the limited research available indicates that the combined value of intercropping is usually higher and the risk is generally lower; however, there is little data on input costs. There are indications, however, that the proportion of land being intercropped is declining. Non-cereal food crops such as beans, cucurbits, cassava and sweet potato are important (and sometimes considered women's crops). They are often insufficiently considered by research and extension and may be underreported in official statistics.

Reducing Labour Demand

Labour is a key constraint to both increased production and the adoption of resource-conserving techniques. Labour bottlenecks at weeding time are a major reason that smallholder crops are generally weeded too late, and not enough times, with significant loss of potential yields. Animal draught has been spreading throughout the region; although this has enabled larger areas to be cultivated, ploughing has often made fields more susceptible to erosion as traditional techniques of leaving trees in fields or using ridges have been abandoned. There have been attempts to introduce animal-drawn cultivators for weeding in most countries in the region, with mixed success (see Box 5.1, Photo 5.5 and Box 11.12).

Sustainable Livestock Technologies

Research indicates that the main constraints to both cattle and goat productivity for smallholders is management and feed rather than the breed; with regard to the conditions normally found among smallholders, indigenous breeds tend to be the most productive and resistant. Post-drought livestock restocking should therefore be done with indigenous animals, and 'improved' breeds should only be introduced in conjunction with changed management practices. A disproportionate amount of livestock research has been on pure breeds of cattle under optimum conditions – there should be a greater concentration on indigenous breeds under real conditions, as well as on goats and donkeys (for draught) and on crop/livestock interactions. More work is needed on indigenous remedies, especially for goats and chickens where the cost per animal of modern medication can be prohibitive.

One proven way of improving management is through supplementary feeding, but many farmers cannot afford it. Mineral licks and protein supplements can be more cost effective than complete feed and there is potential for more home mixing of feed, including the use of poultry

BOX 5.1 THE INTRODUCTION OF THE ANIMAL-DRAWN CULTIVATOR FOR WEEDING IN NORTHERN NAMIBIA

The concept of modern technology as the solution to farmers' problems has been vigorously promoted by both pre- and post-independence politicians. Government tractor hire services, the subsidised sale of donated tractors, and the purchase of tractors by businessmen–farmers have all tended to reinforce the belief that draught animal-powered technology is primitive.

In fact, both economic and ecological considerations dictate that tractor-powered disc ploughing is unsustainable in the case of extensive pearl millet production. Average yields of 200 to 300 kilogrammes per hectare do not justify the cost of tractor ploughing. Discs damage the fragile structure of the light sandy soils, both by compressing it and, when cultivating at speed, pulverising it. Also, with so few tractors in operation relative to demand, farmers have to wait their turn for them to become available. This often causes them to miss short-lived wet conditions necessary for good crop establishment.

On most Northern Communal Area soils, the main function of ploughing is to prevent weed growth rather than to prepare the seedbed. Therefore, the introduction of the animal-drawn cultivator to carry out inter-row weeding has proved a double success. Not only has it facilitated weeding (identified by farmers in most areas as the major factor limiting their production), and thus increased the area that a household is able to cultivate, it has also meant that there is less need to plough in the first weeds. This, in turn, enables farmers to cultivate and sow immediately after the first rains or to dry plant. This is one more way in which farmers can spread risks associated with irregular rainfall. It also allows emerging seedlings to benefit from the nutrient flush associated with the first rains.

It has taken a concerted effort to persuade farmers that draught animal-powered technology is the most suitable for their needs. An extension programme deals with a range of constraints to adoption, and includes:

- on-farm demonstrations of line planting and using the new cultivators for weeding;
- programmes teaching farmers to train draught animals to walk in straight lines when weeding;
- programmes aimed at production of supplementary feeding for draught animals at the end of the dry season;
- the promotion of draught animal-powered transport (presently little used in some areas) throughout the year (this adds value to draught animals over an extended period and hence encourages farmers to target improved feeding programmes at them, rather than at all animals);
- offering farmers choice by introducing two different models of cultivator (the Western Hoe from Senegal, and the Maun from Zimbabwe);
- the unsubsidised sale of cultivators at commercial retail outlets;
- the promotion of local manufacturing of cultivators and spares.

Early signs are that adoption in the north central regions has been rapid. In the Kavango and Caprivi regions, the sale of cultivators has been low in the first year of sale.

Source: EDC, 1997d

Photograph 5.5: *Woman hand cultivating in ridges in Mozambique – animal and tractor ploughing is reducing the use of ridges in some areas, which can increase soil loss from erosion.*

manure as a component. There seems to be potential for the development of fodder banks both for fattening and dairy (see Box 5.2).

Draught animal power is a constraint for many farmers, particularly following recent droughts. Possibilities for alleviating this include:

* improving performance of existing animals through supplementary feeding;
* spreading the ploughing period by more winter ploughing;
* minimum tillage;
* introducing light implements for donkeys and for weeding.

Grazing

There is considerable debate over grazing management systems, with some advocating high-intensity, short-duration grazing (see Box 5.3) and others a four-year rotation: three of grazing and one of rest. Actually, what is needed is a variety of management techniques, depending on the particular needs of different areas at different times; for instance, encouraging the regeneration of trees may require a longer, less frequent rest period than grass regeneration, and the different eating habits of cattle and goats also need to be considered. Therefore, proactive grazing management, based on understanding and observation of the veld and the effect of grazing on it, needs to be researched and encouraged. This differs from the blueprint approach used by some agricultural extension departments in various grazing schemes (see Chapter 8).

The value of browse has generally been underestimated in rangeland extension; it is estimated in Zimbabwe that browse provides 20 per cent of intake and that over 40 per cent of farmers cut branches for browse (Balderrama et al, 1988). Browse can be particularly important in times of drought.

Traditional static analyses of carrying capacity (based on the maximum number of hectares per large stock unit – LSU) are not particularly relevant in drought-prone communal areas because:

- Optimum stocking rates vary according to the management system and the outputs required (optimum rates for ranched prime beef being lower than for communal area meat/draught/milk/savings systems).
- In areas of highly variable rainfall, higher stocking rates can be achieved by allowing livestock to move relatively large distances to areas with better grazing at a particular time.
- In areas with highly variable and cyclical rains, optimum stocking rates change according to the amount of rains, and therefore livestock numbers rise and fall according to the quantity of rain.
- The pattern of grazing (length, timing and intensity) affects grazing and therefore carrying capacity. Short, heavy grazing in which both palatable and unpalatable grasses are eaten, and the soil surface is broken up, followed by a recovery period, is thought by some to cause less damage and makes heavier stocking possible.

Community management of sustainable grazing is discussed further in Chapter 8.

Wild Product Management

A variety of wild products are underrecognised and underresearched components of smallholder livelihoods. It is now better recognised that

BOX 5.2 ICRAF, AGROFORESTRY AND SMALL-SCALE MILK PRODUCTION IN ZIMBABWE

SADC–ICRAF research in Zimbabwe focuses on problems of declining soil fertility and shortages of quality fodder and tree products in collaboration with the Department of Research and Specialist Services (DR&SS). Major activities have been:

- on-station research in improved fallow with *Sesbania*, *Cajanus* and *Acacia*;
- improved biomass transfers (both on-station and on-farm) through planting of leguminous trees on the farm so that foliage, prunings and cuttings can be used as organic manure instead of collecting these from the *miombo* woodlands;
- on-station and on-farm growing of leguminous trees, for use as supplementary feeds in dairy cow diets.

Smallholder milk production is being seen as one route towards improved livelihoods for some farmers in communal areas. The growing of high-feed value leguminous trees is being developed by ICRAF to reduce reliance on bought-in concentrates and to increase sustainable profits.

Juru Milk Centre

The milk centre is the focal point of a project in Juru, encouraging smallholders to take up small-scale dairy production using improved cattle. The centre acts as a buying point for milk from its members. This milk is either processed into sour milk, packed in 500 gramme bags, which are sold to local outlets, or stored in a bulk tank and then sold on to the Dairy Marketing Board. Members are prohibited from selling milk direct to neighbours (which is apparently against the law). The centre also sells inputs to farmers, such as concentrates and medicines.

The centre has a nominal membership of about 300 farmers, although of these only 96 are currently active in producing milk; some of the others are reported to be still in the process of meeting the full requirements of membership – the development of a fenced acre of improved fodder (typically napier grass), the building of a concrete milking shed, and the ownership of a dairy cow. This represents a considerable investment in money and time; in addition there is a joining fee of Z$250 (US$24), and an annual subscription of Z$100 (US$9). This means membership tends to be concentrated among the richer households – especially those with off-farm income. The ownership of the centre is in the process of being transferred to the member farmers.

Profits tend to be low for participating farmers because of the high cost of concentrate in relation to the price they get for their milk. Farmers received Z$2.48 per litre after deductions for the milk centre charges, while concentrates can cost them up to Z$2 per litre of milk – leaving no margin for other expenses or for profits. In this context, ICRAF and DR&SS have been encouraging farmers to grow fodder banks of fast-growing leguminous trees such as *Leucaena leucocephala*, *Calliandra calothyrus* and *Acacia angustissima*. The trees are pruned several times during the growing season and the leaves either fed directly to the cows, or dried and stored in bags as hay. The leaves form a high protein feed, which can be mixed with crushed maize and can replace expensive concentrate.

The agroforestry component of the programme has been in existence for only two years, so it is too early to judge long term sustainability of the system. However, some preliminary points can be noted:

- Tree growth is good in this relatively favourable area. However, production would be reduced in lower rainfall regions more typical of communal areas.
- Labour requirements are high, including the cutting and carrying of both the napier fodder and the tree prunings, and the daily transport of milk to the milk centre.
- Because of the high start-up costs, schemes like this tend to benefit only a minority of relatively better-off farmers.
- Such schemes are very dependent on the efficient management of the dairy centre, so that maximum benefits are passed on to the member farmers.
- Capital costs of the centre are high relative to the number of farmers – wide-scale replication would mean either considerable donor funds being targeted on relatively better-off farmers, or the need for a large loan (and repayment costs would squeeze profits still further).
- More relaxed rules on selling milk direct to neighbours and on the use of local materials in the milk shed might enable the benefits of a project like this to be more widely spread, without necessarily having a detrimental effect on hygiene.

Wider Applicability

There is potential for introducing fodder trees more widely in order to fatten livestock and feed draught animals, particularly in the dry season, in preparation for ploughing. Farmers would have to be convinced that the additional capital investment in fencing and the labour needed in tree establishment and cutting fodder are worth the benefits.

Source: EDC, 1998a

communities manage natural woodlands, selecting which trees to cut and which to conserve. Management becomes more intense where individual rights are easier to protect, such as when trees are close to the homestead. By-laws preventing tree cutting often do not encourage the rational management of trees by farmers, nor do they foster community responsibility. The community management aspects of wild products are discussed in more detail in Chapter 8.

Much of the emphasis during the past 20 years has been on eucalyptus seedling production and the establishment of community woodlots. Evaluations of the limited success of many of these programmes have indicated a number of new directions:

- the decentralisation of large government nurseries to smaller, often community or individually managed nurseries;
- a shift from community woodlots to supporting planting by individual households, commonly near the homestead, or on field boundaries;
- prioritisation by farmers of fruit trees and multipurpose trees – this

Table 5.4: *Technologies for Improved Livestock Sustainability*

Objective	Techniques	Comments
Improved cattle, goats, sheep	**Improved breeds.**	In general, management is the limiting factor and local breeds tend to do better than exotics under typical smallholder conditions. Some programmes are cross breeding goats at small-holder level, but are experiencing high mortality of pure-bred males.
Improved chickens	Improved intermediate-level chicken management.	Much of the development in chicken production has been into high-input, high-output egg and broiler systems. There is, however, potential for improvement to the ubiquitous indigenous chickens found in every household: • supplementary feeding with maggots, termites and dried blood from local livestock slaughter; • brooders to reduce chick mortality from predators; • new heat stable Newcastle disease vaccine; • more work on traditional remedies; • black Austrolopes chicken seems to be able to combine productivity and hardiness, under some smallholder conditions.
Improved animal health	**Vaccines and treatments.**	Attempts to develop cost-effective delivery, through farmer training and para-vet programmes, is at times blocked by professional boundary protection by vets. Some progress is being made on researching and popularising traditional remedies.
Improved nutrition	**Fodder crops** (such as trees and fodder legumes) and zero grazing.	Traditional livestock regimes are typically low input, low output. Growing fodder increases input, particularly of labour, and is often only cost effective if a corresponding increase in output value is achieved. This is possible where enabling environment favours intensification – for instance, when there is a market for milk (see Box 5.2).
	Supplementary feeding.	Supplements such as minerals or locally made concentrates can be cost effective in certain circumstances. Also, supplementary feeding of key breeding stock during drought (Chapter 12).
	Veld reinforcement and rehabilitation.	Seeding highly productive species is effective in high potential areas; however, there is little experience in drier areas under communal management.

		Degraded rangeland can be technically rehabilitated – for instance, by the construction of ridges, infiltration pits, and laying brushwood in gullies. However, the issue is likely to be the benefit in relation to the cost, and the organisation needed to achieve this in common property areas. Prevention is likely to be better than the cure.
	Appropriate and sustainable grazing systems.	Much effort has gone into fenced (rotational) grazing schemes and trying to enforce specific carrying capacities – with very little success. Challenges to this orthodoxy comes from proponents of tracking management and holistic resource management (see Box 5.3).
Diversif-ication	Diversification into new 'appropriate' species.	Many efforts with rabbits and guinea pigs have had limited impact. Guinea fowl is a common species for smallholders to rear in some areas. Ostrich farming is expanding but not thought suitable (so far) for smallholders. Where conditions make it possible, there is some limited success with introducing fish farming, often combined with pig effluent.

has led to a need to increase training in grafting techniques;
- more emphasis on the conservation and management of natural regeneration, rather than just new planting;
- the protection of trees from livestock, which is often a limiting factor;
- the use of micro catchments to water trees, which can be a key ingredient in some locations.

Wild products are often particularly important in times of crisis (eg as famine foods), to women and children and to remote or marginalised groups. There is potential for more research into the development of particular products; in Botswana this has mainly been performed by NGOs, although there are recent signs of greater interest by government and the private sector (see Box 5.4).

There is currently much interest in domesticating some veld product species for on-farm production. While the desire to domesticate veld products is understandable, there are also some very real dangers (see Table 5.5).

WHAT POTENTIAL FOR RESOURCE-CONSERVING TECHNOLOGIES?

What is the potential for technology to enable sustainable intensification of smallholder agriculture in Southern Africa? It is difficult to answer this

Box 5.3 The African Centre for Holistic Resource Management (ACHRM)

Holistic resource management (HRM) is described by proponents as 'a unique, goal-driven decision making process that integrates human values with economic and environmental concerns, resulting in management that is proactive and sound – ecologically, socially and economically'. Holistic resource management is a small international movement, based in the US, but founded by a Zimbabwean, Alan Savory. The movement opened its Africa offices in Harare in 1992.

Holistic resource management is best known because of its application in range management. A small but growing number of commercial ranches in Mexico, the US, Zimbabwe, Namibia and South Africa apply its principles. The system involves heavy impact grazing: very heavy stocking of small areas of range land for a very short period of time. In this way, there is more competition and therefore less selectivity of grazing, and unpalatable species tend to get trampled. The high-impact, short-duration grazing breaks up the top layer of soil and the trampled grass produces a mulch, creating conditions for seed germination and grass regeneration. This type of grazing is considered to be closer to the natural grazing of large herds of antelope and it is surmised that the veld species are adapted to this type of pressure (Savory, 1991; Bingham, 1990).

Proponents of high-impact, short-period grazing claim both considerable increases in carrying capacity and increases in veld biodiversity. So far, there are limited objective data to verify this, but the Zimbabwean DR&SS are currently planning an evaluation of the system. To date, holistic resource management has mainly been taken up by commercial farmers. However, since December 1994 the ACHRM has started to work with communal farmers in the very dry Matetsi communal lands of Hwange, in Zimbabwe, using the concepts and practices of holistic resource management. During the initial two years, farmers were selected and trained to become trainers themselves. The implementation phase started in seven villages with the trainers, who operate in pairs, composing their own support group of 40 to 50 persons. Holistic goals are formulated at community level, and disagreements and contradictions are made explicit. After formulating holistic goals, plans for action are agreed upon.

It is too early to objectively evaluate sustainable impact, although the centre claims that communal high-impact grazing is already showing promising results.

Comments and Lessons Learned

- HRM is important in that it highlights, once again, that the type of grazing pressure is every bit as important as the static stocking rate formulas of X hectares per LSU. Despite much evidence to the contrary, static stocking rates are still used in range management throughout the region. It is crucial to understand that undergrazing or prolonged light grazing are just as important problems as overgrazing.
- Extravagant claims are being made about the benefits of HRM. Meanwhile the agricultural mainstream has remained unconvinced. The evaluation by DR&SS is welcomed, both in order to get some rigorous data, and as a way of breaching the HRM/establishment divide.
- Lessons learned more generally about the diversity of smallholder agriculture and its environments must not be forgotten; a type of grazing scheme suitable for one area is not necessarily ideal for another.
- Proponents of HRM claim that the decision-making process is unique – with a number of questions used to test whether decisions lead towards the holistic goal. However, in practice, it is difficult to see how it is qualitatively different from other techniques of participatory planning. Claims to uniqueness do not encourage open sharing with others.

Source: FDC, 1998a

BOX 5.4 DEVELOPING VELD PRODUCTS IN BOTSWANA

There is increasing recognition in Botswana of the potential for veld products to contribute to rural livelihoods:

- Veld products are important to the poorest and most remote households, who have been the most difficult to assist through conventional programmes.
- Veld products are seen as key components of community-based, natural resource management (CBNRM) schemes (see Chapter 8), and the creation of value for these products is seen as a way to promote their conservation.
- Veld products have the potential for greater profitability and lower risk in some circumstances than crop and livestock agriculture.
- Markets for veld products such as morula, mopane, herbal teas, grapple (*Sengaparile* or *Harpagophytum procumbens*) and truffles are being developed in South Africa, Europe and the Far East.
- Veld products are already an important and unrecognised part of the informal economy and household livelihoods; in addition to the specific products with cash earning potential, households have always used wild vegetables, fruits, firewood, thatch and poles.

Two NGOs – Thusano Lefatsheng and Veld Products Research – have an existing track record of working with communities on the conservation, collection, processing and marketing of veld products. Mainstream economists and policy-makers are now taking veld products seriously and larger-scale production programmes are being planned. For instance, an ambitious programme to collect and process morula (*Sclerocarya birrea*) in the north of Botswana plans to market 12,000 tonnes per year to create 600 jobs and to inject P1.1 million into the local economy.

Source: EDC, 1997a

TABLE 5.5 ADVANTAGES AND RISKS OF DOMESTICATING VELD PRODUCTS

Advantages	Risks
Reduced pressure on fragile and scarce natural resources.	Improved cultivars (with narrower genetic base) may increase yield in an average year
Increased production per plant and per unit area with improved cultivars and improved cultivation.	but destroy those attributes that make veld products particularly important for food security: drought and pest resistance, etc.
Increased production per household because of reduced time spent harvesting from dispersed plants.	Cultivation may destroy the socio-economic characteristics that make veld products so
Care taken over plants as private rather than common pool resource.	important for the poorest remote area dwellers. Where there is a limited market, successful cultivation may be concentrated in less remote, more favourable situations and destroy the market for those in remote areas.

question because so much of existing experience is biased against sustainability. However, we do know that:

• Technology alone is unlikely to achieve very much; it needs to be combined with supporting local institutions and an enabling external environment.
• Single technologies are likely to have limited impact, while combinations are more likely to be successful.

Despite research on individual technologies, there are surprisingly few studies in the region on the impact of combinations of technologies. One of the areas in Southern Africa most challenged by a lack of sustainability is southern Malawi. Here high population and low land availability force smallholders to cultivate small farms under almost continuous maize. These smallholders have little access to organic fertilizer and cannot afford inorganic fertilizer, and so are caught in a downward spiral of reduced fertility and diminished yields. Table 5.6 tries to assess the likely impact of different technologies on increasing sustainability in Southern Malawi.

For southern Malawi, adopting a combination of these resource-conserving technologies would lead, at a conservative estimate, to:

• stabilisation of the farming system;
• improvements in diet quality;
• 50 per cent increase in maize yields;
• addition of a cash crop and some firewood or poles.

To achieve this would require a modest increase in labour demand but few other costs. However, even a modest increase in labour during peak agricultural activities may be impossible for the poorest families to achieve – any spare labour is hired out to neighbours for essential cash or food. The capacity of farmers to adopt a long term perspective is also required, since some of the returns to the increased labour would only be felt after a number of years (for instance, improved fallows would result in a reduction in production in the first two years – a serious constraint for the poorest households).

Therefore, despite being technically feasible, adopting these technologies and achieving sustainable intensification may not happen because of the constraints facing the poorest smallholders. More work will be needed to address some of these constraints in order to create a suitable enabling environment.

CONCLUSION AND RECOMMENDATIONS

A variety of potential resource-conserving technologies exist, combinations of which could make a major contribution to increased sustainability and productivity. More research (often with a longer time perspective), partici-

Table 5.6: *Probable Impacts of Some Best Bet Technologies for Malawi*

Technology	Yield Increases Reported	Best Bet Impact Estimate	Comments
Faidherbia albida	40 to 270% increase in maize yield	20 to 50% on surrounding maize	Many estimates on increased yield are quite difficult to interpret since they compare yields close to and far from mature *Faidherbia* trees. Others come from yields under natural stands of *Faidherbia*, but it is difficult to select a control site. The best bet estimate assumes a planted but imperfect (in terms of density and spacing) stand of *Faidherbia*.
Vetiver grass strips orientated by A-frame level	Stabilisation	Stabilisation and 5 to 10% loss in crop area	These technologies are necessary on sloping sites to prevent continuing erosion and yield declines in addition to whatever other technologies are used.
Rotation/ intercropping with Magoye soya and other legumes	Some estimates give gross margins for soya at eight times, and groundnuts at four times, unfertilized local maize	40% on food quantity and quality	20% due to improved food quality and 20% due to yield in subsequent maize crop.
Improved fallows	Two-year fallows in Zambia increase net present value by 90%; others show 0 to 150% yield increase averaged over four to six years	20% maize yield increase averaged over whole rotation	Costs of land taken out of production felt in year one; benefits only start to be felt in year three and become positive in year four or five. Some additional benefits from poles and fodder.
Relay cropping – *Tephrosia*, etc	0 to 130% increase and doubling of net present value	10 to 30% increased maize yield	More work needs to be done on establishment and more measurements are needed on effects on yields.
Diversification, eg burley tobacco or other cash crops	Burley tobacco has an estimated gross margin of 9205 Kw/ha or 19 times that of unfertilised local maize	10 to 50% increase in value of total farm production	Assumes small farmers are only planting 10% of their farm and that margins will fall with increased national production. Burley requires more labour than maize.

patory technology development and local adaptation are all needed to further develop technologies. However, technical potential alone is not enough:

- The methodology of spread is crucial – scientists love things clear, but the process of facilitating technology development, adaptation and spread at community level is 'messy'. It is, in fact, not a *science* but an art (see Chapter 10).
- There are very real barriers to adopting some technologies, and ways need to be found to reduce these barriers, or alternative techniques need to be developed that are more suited to smallholder needs. This requires affirmative action by research and extension.
- Some techniques need action and adoption at a community level (this is discussed further in Part III).
- Some techniques necessitate a more stable and enabling environment to provide an incentive for farmers to take a long term view (this is discussed further in Part IV).

Part III

Sustaining Local Institutions

Chapter 6

Why Local Organisation Is Important

INTRODUCTION

An organisation is defined as 'a group of individuals bound together by some common purpose' (North, 1990). Local organisations, which are primarily under the control of local farmers and communities, are necessary for facilitating sustainable agriculture for a number of reasons:

- Services are mediated through local organisation to lower transaction costs (eg a credit organisation may work through local groups because it is cheaper and more effective).
- Local organisation is the basis for the network of entitlements and obligations that is fundamental to the operation of smallholder agriculture (eg the sharing of draught animals for ploughing, reciprocal exchange of labour, division of livestock herds between *kraals* of relatives and neighbours). Local organisations often define the household's use rights to resources (eg membership of a community, perhaps with allegiance to a specific headman, may give rights to land and grazing).
- Organisations influence norms of behaviour that can be a crucial but elusive factor in sustainability (eg peer group pressure may influence a household's practices in cutting trees and controlling livestock).
- Local organisations are (or can be) responsible for managing common pool resources (see Chapter 8).
- Organisations can be a means of advocacy to improve the policy environment in favour of the members (see Chapter 7).
- Local organisation is needed to undertake tasks which are too large for an individual or household (eg the construction of a small dam, catchment erosion control or management of a group vegetable garden); see Photo 6.1.

Photograph 6.1: *Some tasks, like erosion on this scale which crosses the fields of several different households and silts the dam used by the whole community, needs to be tackled at the community level. Community organisation is needed to achieve this.*

Lack of effective local organisation is increasingly recognised as limiting the sustainability of development. A study four to ten years after completion of 25 World Bank-financed agricultural projects found that continued success was associated clearly with local institution building (Cernea, 1987). It has been suggested that rather than viewing natural resource availability as the limiting factor, it is more appropriate and more positive to regard effective institutions as the true scarce resource.

While the importance of local organisation needs to be emphasised, it should not be idealised. Local organisation:

- is not necessarily democratic (although this begs the question of defining democracy!);
- can be a means of expressing ethnic or other rivalry or exclusion;
- may be exploited by powerful individuals for their own ends;
- may exploit or oppress women, youth or other less powerful sectors of society.

Therefore, the *quality* as well as the *quantity* of organisation is important.

HOUSEHOLDS, ORGANISATION AND SERVICES

The relationship between individual, household, local organisation and agricultural services is illustrated in a simplified form in Figure 6.1. The

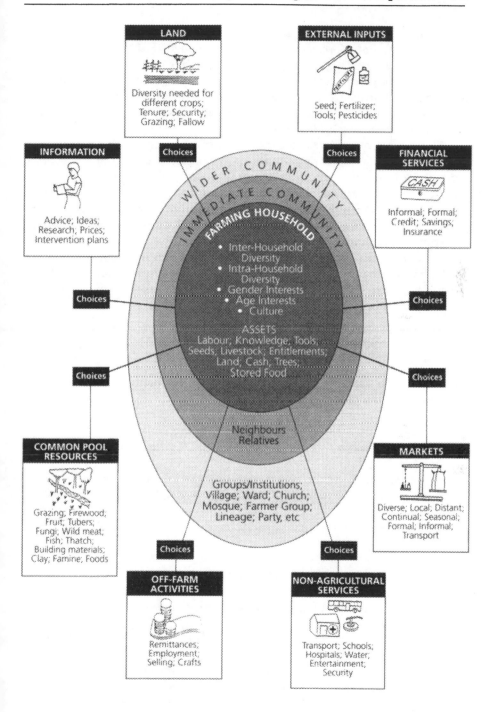

Figure 6.1: *The Farming Household Context*

household, defined here as an income sharing unit, is not necessarily discrete – different sources of income tend to be shared between different individuals. Within the household, each member has different interests, often depending on gender or age, and culture plays a role in determining how these interests are expressed. The household has a range of assets, including the more intangible ones such as entitlements and knowledge, which are used to secure livelihood.

The household relates to services and resources both directly and through various layers of organisation. This organisation is not necessarily formal (for example, with discrete membership and committees), but can be part of the informal network of extended family, neighbours and lineage. Often the informal networks are underestimated by outsiders working for development organisations – both their positive and their constraining potential. There is growing recognition that, despite efforts to create new local organisations, such as village development committees in Zimbabwe, traditional organisations of *kraalheads* and chiefs still have greater legitimacy, particularly in relation to natural resource management, and may be more accessible to women (EDC, 1998a).

PARTICIPATION

Participation is increasingly seen as a panacea for many of the intractable problems of rural development, such as:

- providing sustainability when project funding ends;
- making agricultural research, extension and policy relevant to smallholders;
- management of common pool resources (see Chapter 8);
- overcoming high transaction costs of providing services to smallholders;
- trying to ensure that local capacity is organised in the interests of the poor.

However, in practice the situation is more complicated, with participation meaning different things to different people (see Table 6.1), and very often participation can only be part of a solution that requires changes on a number of fronts.

During the colonial period and after independence, strongly centralised governments ignored, oppressed or coopted local organisation. For example, woodland management became the responsibility of forest departments using authoritarian regulations prohibiting tree cutting, which often had the opposite effect from that intended – proving a disincentive to local tree planting and management. This approach is generally giving way to a more plural or participative system in which the importance of community involvement is at least acknowledged. In reality, it is more complex:

Table 6.1: *A Typology of Participation: How People Participate in Development Programmes and Projects*

Typology	Characteristics of Each Type
Passive participation	People participate by being told what is going to happen or has already happened. This is a unilateral announcement by an administration or project management without listening to people's responses. The information being shared belongs only to external professionals.
Participation in information giving	People participate by answering questions posed by extractive researchers using questionnaire surveys or similar approaches. People do not have the opportunity to influence proceedings, since the findings are neither shared nor checked for accuracy.
Participation by consultation	People participate by being consulted and external agents listen to their views. These external agents define both problems and solutions, and modify these in light of responses. Such a consultative process does not concede any share in decision-making and professionals are under no obligation to take on board people's views.
Participation for material incentives	People participate by providing resources – for example, labour – in return for food, cash or other material incentives. Much on-farm research falls into this category, since farmers provide the fields but are not involved in experimentation or the process of learning. It is very common to view this as participation, yet people have no stake in prolonging activities when the incentives end.
Functional participation	People participate by forming groups to meet predetermined objectives related to the project, which can involve the development or promotion of externally initiated social organisation. Such involvement does not tend to be at early stages of project cycles or planning, but rather after major decisions have been made. These institutions tend to be dependent upon external initiators and facilitators, but may become self-dependent.
Interactive participation	People participate in joint analysis, which leads to action plans and the formation of new local institutions or the strengthening of existing ones. It tends to involve interdisciplinary methodologies that seek multiple perspectives and make use of systematic and structured learning processes. These groups take control over local decisions and people have a stake in maintaining structures and practices.
Self-mobilisation	People participate by taking initiatives independently of external institutions in order to change systems. They develop contacts with external institutions for the resources and technical advice they need, but retain control over how resources are used. Such self-initiated mobilisation and collective action may or may not challenge existing inequitable distributions of wealth and power.

Source: Pretty, 1994; adapted from Adnan et al, 1992

- Regulations and local enforcement practice often remain authoritarian, although this is changing.
- Local officials often do not have the skills or motivation to move from an authoritarian to facilitatory approach. In Zimbabwe, as decentralised rural district councils have taken on greater responsibility, one of their first responses to deforestation has been enforcing by-laws by fining people for cutting trees, while at the same time recognising that this approach is unsustainable (EDC, 1998a).
- Efforts to involve the local community are often driven as much by a need to save money as by a commitment to participation.
- Local organisation is not yet sufficiently strong to take over the role of the retreating state. This has tended to leave a vacuum, with common pool resources, such as firewood and grazing, becoming unregulated, open-access resources, which are not being used sustainably; and levels of theft, often accompanied by violence, severely hampering agriculture, particularly the keeping of livestock and the growing of high-value crops.

Participatory techniques are increasingly being used by both government departments and NGOs throughout the region to plan development interventions. These are most commonly referred to as participatory rural appraisal (PRA), although, in practice, the level of ownership by the community is often quite limited, making them closer to rapid rural appraisal (RRA); see Table 6.2 and Photo 6.2.

For development programmes to be genuinely participative, local communities need to be involved in setting the objectives and designing the programme, not just adding detail to predetermined project initiatives. This can be a long term process, with growing participation of the local communities and project methodology evolving with experience. Lessons for donors are:

- Long term commitments to an area and community may be needed.
- Participatory evolution of a programme may require greater flexibility, for instance in future outputs, than is commonly encouraged by the logical frameworks currently demanded by many donors.

In practice, community participation does not usually mean that the views of all the different members of the community are heard, or taken equally seriously. Affirmative action is often necessary to ensure the voice of the marginalised and issues of sustainability are heard. For example, although participation by women has been on the development agenda of most organisations and governments in the region for more than 15 years, it has been difficult to achieve in practice. For instance, an evaluation of gender issues in relation to community participation in six natural resource management (NRM) projects in Malawi (CURE, 1996) concluded that:

Table 6.2: *Comparing PRA and RRA*

Criteria	PRA*	RRA*
Objective	The community decides on development priorities and plans (which may subsequently be presented to a government or agency for support).	The agency decides on relief or development priorities and plans.
Timescale	The appraisal can be short or prolonged, but it is part of a longer term process in the community.	The appraisal is normally relatively rapid, but may be part of a longer term data gathering and planning process within the agency. The appraisal is also part of the events affecting the community and has an impact, whether or not this is planned by the agency.
Key actors	Community members, often facilitated by outsiders.	Outsiders, often facilitated by community members.
Interpret-ation of results	By the community.	By outsiders.
Techniques used	A wide variety – mapping and modelling, time lines and change analysis, seasonal calendars, daily time-use analysis,institutional diagramming, linkage diagramming, matrix scoring, well-being ranking, participatory planning – can share techniques with RRA.	A wide variety – direct observation, semi-structured interviews, transect walks, mapping, key informant interviews – can share techniques with PRA.
Political correctness	High (sometimes there is no funding without it).	Moderate (sometimes considered superseded by PRA; however, the objectives are different and RRA can be more appropriate in particular circumstances – see Whiteside, 1997)
Usefulness	Depends on context and attitudes of those involved.	Depends on context.

* In reality many appraisals combine elements of both approaches.

- **Women's participation (in terms of quality rather than quantity) is poor in decision-making but high in project implementation activities (eg the hard work).**
- **Systems of inheritance (matrilineal or patrilineal) have an important effect on gender participation in NRM activities.**
- **Despite NGO desires to see gender equality in project outcomes, few deliberate strategies were put in place to address gender issues in NRM. Instead, it was observed that the existing gender dynamics**

Photograph 6.2: *PRA techniques can help communities to decide for themselves what their priorities are and how to handle them – here members are 'voting' by putting different numbers of beans onto pictures of the different problems facing the village. But how deep is the commitment to participation among different professionals?*

within the community, and the literacy levels of the women, were more likely to determine gender equality within the projects.

• Improving gender relations at a community level can have a positive impact on project success.

• There is a need for project staff to be exposed to skills which can be used to address gender issues in NRM.

A major barrier to introducing more participatory work has been a short-age of personnel, particularly field staff, who combine agricultural knowledge with skills in, and commitment to, participation. This has been recognised and is starting to be addressed; in Zimbabwe and elsewhere, NGO-run Training for Transformation short courses have been used by a wide variety of organisations over a number of years and are well regarded. The Participatory Ecological Landuse Management network (PELUM), with its PELUM College, is an attempt to address this skills shortage on a regional basis (see Box 6.1).

Often programmes committed to participation find themselves chang-ing during the life of the programme as the understanding of and by the local community develops. The ActionAid programme in Dowa (see Box 6.2) is not unusual in its evolution – from having an emphasis on a specific project intervention, to an increased concentration on strengthening local organisation. The realisation that, as a programme changes, this may mean changes in management structure and staff capacity is also common.

> ## Box 6.1 PELUM Association
>
> PELUM, which stands for Participatory, Ecological Landuse Management, has a regional membership, bringing together organisations in East and Southern Africa. PELUM developed out of a common need of local development organisations for improved training and resources for their field workers. There is a great shortage in the region of field workers and senior staff with understanding and experience of both participatory approaches to development and sustainable agricultural methods. Members set the priority of the association as facilitating learning and networking, by linking the members' experiences, methods and approaches. Working groups have been set up in various countries, with a regional desk in Harare.
>
> The formal launch of the association was in October 1995. Activities to date have been:
>
> - Developing and distributing training materials.
> - Establishing short courses, based on analysing the training needs of members, and with a focus on training to strengthen local trainers.
> - Developing longer term training in a PELUM College. Initially, a two-year sandwich course is being developed, allowing students to travel between 14 different organisations within Zimbabwe, therefore being dubbed the 'college without walls'. These 14 organisations will provide training in sustainable landuse and community development, with an aim of providing students with a well-rounded, long term training. The process of setting up the college has brought together universities, national NGOs, community-based organisations and government departments, in a unique alliance to offer comprehensive training in sustainable development.
>
> *Source:* EDC, 1998a

Limits to Organisation and Participation

Participation and organisation are important but caution is also needed:

- Participation has costs, as well as benefits, for the community – typically these are in time spent in meetings and participating in activities. Often these costs are particularly heavy for the busiest and poorest members of the community, such as women. Benefits of participation need to substantially outweigh the costs for an activity to be sustainable.
- For the poor, participation is not always the 'ideal' promoted by some development practitioners, but a 'necessary evil'. For instance, most households would prefer to be able to afford to pay for tapped water like the rich, rather than participate in water committees and help dig the well. Similarly, most would like to be able to get a loan straight from the bank like the rich, rather than form mutual liability groups with neighbours and relatives in order to qualify for credit (Chapter 11).

Box 6.2 ActionAid Programme, Dowa, Malawi

In 1991, ActionAid started its Msakambewa Rural Development Area Project in Dowa district. The project area is in Central region, has a population of 38,000, 36 per cent of farms are less that one hectare, and of these the average farm size is 0.66 hectares. 80 per cent of farms grow maize only and manage to produce only 57 per cent of necessary calories; average cash income in 1991 was 154 Kw (US$45).

The project was planned to cover three sectors – health and sanitation, agriculture and education. Initially, the project was implemented on a sectoral basis; however, this approach was not conducive to integrating diverse development activities at village level, and the project was reorganised along geographical lines. The agricultural component was to be implemented in coordination with the Ministry of Agriculture.

The core of the project was 'to find ways to bring smallholder farmers (<one hectare) across the line from below subsistence to cash surplus, through improved yields of maize and diversification into other crops'. The aim was to assist communities in determining their own development priorities and to plan and implement these with the assistance of ActionAid field staff.

The agricultural component has concentrated on a number of interventions:

- **Land husbandry:** this involves training and motivating village-level land husbandry volunteers to use A-frame levels to align contour ridges.

	1992	1993	1994
Number of land husbandry volunteers	49		136
Number of hectares aligned	366	257	559
Number of households	356	430	933

Although the quantity of the output was considered satisfactory, ActionAid were concerned about the lack of spontaneous contour alignment by villagers, suggesting little conviction on behalf of villagers that contour alignment was worth the hard work involved – or a lack of skills transfer by the land husbandry volunteers.

- **Agroforestry:** this mainly involves nurseries, established by interested farmers rather than at headman level as originally planned. Around 12,000 seedlings were distributed per year, with a survival rate of 47 per cent. The main emphasis has been on *Faidherbia albida* and alley cropping species; less progress has been made with live fencing and encouraging natural regeneration.
- **Food security:** the main emphasis has been on seed loans, particularly field beans and soya beans, with repayment in kind (with 20 per cent extra as interest). This programme has been popular with the farmers and repayment rates have been reasonable. An attempt to encourage communities to keep their own village seed stores was introduced. However, the perceived risk of pooling seed has meant that these have been more often used as multipurpose community centres than as seed banks.

- **Credit groups:** these groups are formed at village level, and credit is used to buy inputs such as seed and fertilizer. Credit groups have a saving component and need to operate successfully for six months before getting a loan through ActionAid. Originally, ActionAid provided the loan straight to the credit group, but more recently ActionAid has been channelling the funds through the local village development committee to increase local control and ownership. Interest rate is 18 per cent and repayment rates have varied between committees – only committees with a high repayment rate can get a new loan the following year.
- **Income-generating activities:** credit has been given to groups to undertake a variety of income-generating activities, including egg production, goats, bee-keeping, vegetable gardening, fish ponds and rabbits. A review by ActionAid revealed that there was a large variation in success rates between enterprises, with a need for better economic analysis of viability and stronger technical back-up in some cases.
- **Farmer participatory research:** this was done in conjunction with the local Ministry of Agriculture. Initially the trials had been rather top down, based on researcher priorities and researcher methodologies; however, there is interest in developing a more participatory approach, with farmers and researchers jointly working on ideas to overcome problems identified by farmers. The trials were considered to have had a useful demonstration effect for new crops.
- **Village access/feeder roads:** the objective has been to enable village access to basic facilities, such as hospitals, schools and markets, by constructing stream crossings and bridges. A review in 1995 noted: 'of all the activities this seems to be the only one that is truly community-led... it has been very successful, which shows that communities do know best!'

This programme is particularly interesting: it has evolved through trying to find the most appropriate and sustainable way to work with farmers – this has been partly due to ActionAid's own experience and partly due to national political changes with the consequent promotion of different local organisations, principally village development committees (VDCs). Changes include the following:

- There is greater emphasis on strengthening VDCs and on trying to implement programmes through them. Therefore there is greater recognition of the need for institution building as a means to implement technological change.
- This has had consequences for ActionAid field staff, requiring smaller numbers of more skilled staff to work with committees, rather than directly with farmers. The field supervisors are being moved from the central project office to live and work from the villages.

Source: EDC, 1997b

- Sometimes supporting private entrepreneurs rather than community group formation may be more sustainable, more cost effective for the community and/or more cost effective for the development agency. Therefore, the choice of whether to support community group development or private entrepreneurs, or a combination of the two, needs to be based on reason and local possibilities rather than predetermined ideology.

- Existing informal linkages (often nearly invisible to the outsider) may be more important than structures set up by a development agency; local understanding is needed before leaping into 'institution building'.
- Being committed to participation does not imply that local communities are always right. This is not the case; although I would, however, argue that, right or wrong, communities have a right to decide on issues that affect them. Best results from participatory working often come when members of the community, with their intimate local knowledge, interact creatively with external professionals, who have a different knowledge base – together they create a 'whole that is greater than the sum of the parts'. Achieving creative synergy between local communities and professionals is an important attribute that few projects achieve.

CONCLUSION AND RECOMMENDATIONS

Local organisation is important to:

- lower transaction costs for services and to link with outsiders;
- mediate entitlements for household resource use;
- influence norms of behaviour;
- manage common pool resources;
- advocate for improved policies;
- undertake large tasks.

Strengthening local organisation is often as effective and more sustaining than other more concrete activities of development programmes. Informal local organisation, involving relatives and neighbours, is often an underestimated factor in development. Although participation is increasingly recognised as a vital component of development, there are many different types and levels, and there are also some costs. Therefore, agencies need to understand the local situation and to consider various options before starting on group formation.

Development programmes often use participatory techniques without having a fundamental participatory approach, limiting ownership by the community in the overall process. Participatory development is neither a short term process nor one where the outcome can be defined in advance; this means facilitators and donors require:

- long term commitment to an area and a community;
- more flexibility than is often encouraged by the use of logical frameworks.

Participation is both an approach and a process; the right attitudes are needed and development programmes need to be able to evolve to meet the developing demands of the community. Field workers must combine technical agricultural knowledge and the experience to work participatorily with local communities.

Chapter 7

Farmer Groups and Unions

INTRODUCTION

Farmer groups, associations cooperatives and unions can have a number of roles:

* task orientated (eg building a dam or CBNRM – see Chapter 8);
* linkage orientated (eg to service providers);
* service provision orientated (eg providing services, such as inputs or marketing themselves);
* advocacy orientated (trying to change the policy environment).

In practice, many groups and unions combine a number of these roles.

GROUPS LINKING WITH SERVICE PROVIDERS

Farmers' groups are used by their members and by service providers (whether commercial, government or NGO) throughout Southern Africa as a way of lowering transaction costs. For a service provider or development organisation deciding to work through farmers' groups, there are a range of points to consider:

* *To work through new groups or strengthen existing groups?* Building on groups that already exist has some advantages, but specially formed groups can sometimes be more appropriate to a particular task. Several countries in Southern Africa have experienced a proliferation of different village-level groups set up by various agencies, often involving the same few active individuals. Clearly, a more coordinated or multitask structure becomes necessary in such circumstances.

- *Size of group and number of tiers?* Larger groups have lower transaction costs for the service provider; smaller groups have greater cohesion between the members (15 to 25 is sometimes considered a manageable maximum). In Zimbabwe, the credit groups set up by the Agricultural Finance Corporation are considered by some to be too large, resulting in insufficient group cohesion and management difficulties. Groups can be structured in a number of tiers (eg local group, village development committee and ward development committee) – the appropriate tier of interaction depending on the nature of the intervention (see Figure 7.1).
- *Leadership incentives?* There are a whole range of pros and cons relating to whether key community members (such as community-based extension workers) should be rewarded or paid for their work. Sustainability and genuine participation sometimes need to be traded off against getting results. However, what works in the short term may not be sustainable over longer periods. Inappropriate compensation can cause jealousy or biases. Coordination may be needed between different organisations working in the same area to avoid discord. Often it is more appropriate and sustainable for the community to decide and remunerate the person, rather than the outside agency.
- *Role of existing institutions and leaders?* Some projects try to work with and strengthen existing traditional or political structures. Others, concerned about lack of participation, elitism or gender imbalance in the existing structures, tend to by-pass them. In some projects there is compromise, with the new structure existing as a subcommittee of the old or with the headman as an ex-oficio member. In one case in Zambia, the headmen were excluded from groups in order to act as impartial arbitrators in case of disputes (EDC, 1997c). In Zimbabwe, the traditional structure of *kraalhead* and chief tends to have more legitimacy and authority in dealing with issues of land and natural resource management than the newer village development committees (VIDCOs), and are also reported to be more sympathetic to women (Rukuni et Al, 1994).
- *Affirmative action for the less powerful?* There are trade-offs between letting the community decide who to represent them and insisting on quotas for women and youth. A compromise can be to arrange training or a discussion about representing women and the marginalised, before letting the community decide for itself on how to be represented.
- *Permanent or temporary?* Groups can form around a specific task, accomplish it and disband, or be of a permanent nature. The capacity building associated with the formation of temporary groups is often underrecognised. The capacity to organise if the need arises can be as important as having an existing group.

An example of a farmer organisation, set up specifically by an NGO to lower transaction costs in the prevention of animal disease, is the Namwala Cattle Clubs (see Box 7.1).

BOX 7.1 CATTLE CLUBS IN NAMWALA, ZAMBIA

Namwala district lies in an isolated position 400 kilometres west of Lusaka. Population density is low (the 1990 census figure is 83,075 = 3.8 individuals per square kilometre) and cattle rearing has traditionally been central to the Ila economy and culture, with large herds grazed on the seasonally flooded grassland of the Kafue flats during the dry season, and moving south into the bush during the wet season. The total cattle population is estimated to have exceeded 200,000 during the 1970s – four times the human population. However, this has fallen sharply, partially due to declining fertility of the Kafue plains after seasonal flooding was reduced by construction of the Itezhi Itezhi dam. The 1990–91 smallholder post-harvest survey registered 108,643 cattle, including 19,649 trained oxen; these figures tumbled to just 46,818 and 8200 respectively in the following year after the drought and due to disease(GRZ/CSO, 1994).

Cattle in the district are susceptible to a wide range of ailments, including heartwater; trypanosomiasis; lumpy skin virus; tuberculosis; haemorrhagic septicaemia; black quarter; brucellosis; foot rot; eye infections, and foot and mouth (GRZ/MAFF, 1994). However, the most serious cause of recent mortality is widely recognised to have been corridor disease (locally known as *denkete*), which is a form of east coast fever caused by the organism *Theilleria parva parva*. This is a tick-borne disease, which commercial farmers are able to avoid (at a high cost) through weekly dipping of their cattle. Free dipping was provided in parts of the district by the Department of Veterinary and Tsetse Control Services (DVTCS) up until the 1970s, but was never fully effective and has recently broken down almost completely due to lack of funds (Harvest Help, 1994). Research is also currently being undertaken, with support from the Belgian government, into the development of a suitable vaccine. Meanwhile, the best hope is that continued exposure to the disease should lead eventually to endemic stability within the cattle population, and that mortality can at least be reduced through selective spraying against ticks, combined with other measures (such as controlled burning) to reduce tick populations (Woodford, 1995: 1; Harvest Help, 1994b).

The Namwala Cattle Project (NCP) was launched in December 1993 as a six-month action-research project to explore the scope for promoting selective spraying during the rainy season through three pilot cattle clubs. It was conceived by the Harvest Help representative in Zambia and the district veterinary officer (DVO), and run by a joint management committee of Harvest Help and DVTCS. Involvement of the latter was seen by Harvest Help as essential, both as a source of technical know-how and to ensure that any lessons learned were incorporated into mainstream livestock services. Each club was required to construct its own crush pen out of local materials, and then supplied with spraying equipment along with an initial consignment of acaricide. Further supplies would be financed through user fees paid by the club members.

Despite initial suspicion (arising in part from bad experiences in the past with communal dip tanks), two of the three pilot clubs held spraying sessions regularly through the season; this was judged sufficiently encouraging by Harvest Help to justify funding a three-year follow-up project with the same underlying goals, and a target of establishing a total of 15 cattle clubs in the district (Harvest Help, 1994b). The pilot programme revealed the need for more intensive training, and so the new project provided recruitment and adequate resourcing of an extension worker, over and above existing DVTCS staff in the district. In addition to supporting the clubs, this person was expected to coordinate the training of community livestock workers to be selected by each club.

1994/95 was another drought year, and this constrained farmers' ability to pay for spraying. Clubs also encountered difficulties in establishing a viable pricing system for covering costs; nevertheless, three clubs (including the two successful pilot clubs) succeeded in holding spraying sessions approximately every two weeks. In the 1995/96 wet season, eight clubs were established and seven of these sprayed at least fortnightly. Several clubs also developed a system for permitting members with temporary cash shortages to spray on credit.

The first seven community livestock workers (CLWs) received training in August 1995, and six of them became eligible for access to a 50,000 Kwacha revolving loan facility to finance drugs and equipment. This scheme worked well, and was supplemented by efforts to encourage local traders in Namwala town to carry stocks of popular items. However, distribution of acaricide to clubs continued to be arranged through project staff. Bainbridge (1996) reports that total veterinary drug sales through CLWs, NCP and local traders between August 1995 and July 1996 was 4.4 million Kwacha (approximately UK£2,500).

During the remainder of 1996, there have been at least two important further developments. Firstly, while three of the original clubs ceased to operate, an additional 12 were established, bringing the total number of active clubs to 17. Most of the new clubs are in the south of the district around Chikanta, and much of the enthusiasm for them appears to be originating from immigrant (Tonga and Ndebele) farmers, convinced by the success of the two existing clubs in their vicinity. The second development was the demise of collaboration between government and Harvest Help staff (renamed Hodi at the beginning of the year). Although precise figures are not available, it appears that funds available under NCP (estimated to cost about one million Kwacha) were substantially larger than those available to the DVO from the government (three million Kwacha received for the year against a budget of nine million Kwacha). It was therefore a source of frustration to be unable to divert project resources into activities that were perceived to be a priority – a problem compounded by the relative success of the project in the most distant part of the district (and most expensive to visit).

Therefore, as the end of the second phase of the project approaches, and despite evidence that the underlying concept of selective spraying through cattle clubs can work, its future is in doubt. A two-day meeting of 29 representatives of the 17 clubs in October 1996 focused largely on this question. One proposal was to form an umbrella association to which each club would pay an affiliation fee that would go towards meeting joint costs of chemical distribution and training. A committee was formed, and an affiliation fee for each club agreed. But these fees do not even cover NCP's costs for one month, and management of an association looks problematic given the large distances between clubs – particularly between the southern and northern parts of the district. Moreover, a significant increase in the level of support that DVTCS might be willing and able to offer the clubs looks equally unlikely.

Some CLWs might be able to continue to operate successfully if they make arrangements for the supply of drugs through private traders though they will be handicapped by their limited technical knowledge and by legal restrictions on the treatment they are allowed to provide. Other clubs (particularly in the southern part of the district) may also be able to survive by expanding their range of activities to embrace additional activities, including facilitating crop input distribution and marketing. The chances of both succeeding probably depend on further external funding and technical support.

Issues

- Although farmers' organisations can lower transaction costs and are perceived by farmers to bring benefits, this does not necessarily guarantee sustainability on the withdrawal of external support.
- Perceived benefits of organisation, and farmer capacity, need to be high enough to overcome the increased problems and local contributions likely to be needed after funding and other support have ceased.
- The external environment is also crucial to sustainability – in this case, a supportive and adequately resourced veterinary department, an effective retail network and transport infrastructure could provide the conditions for sustainability, but are currently lacking.

Source: EDC, 1997c

There are many different models of farmers' organisations capable of linking to specific project partners and to outside service providers, or of providing services to members. Circumstances, rather than a blueprint approach, need to determine what is appropriate. Figure 7.1 illustrates diagramatically nine different models from Zambia. Some points to note are:

- the difference between temporary project structures and community organisation intended to be sustainable;
- how permanent, non-project institutions, such as retailers and government departments, can link with community organisation fostered by the development project;
- the diverse variations in hierarchies of community groups and the links between the different levels.

ASSOCIATIONS AS SERVICE PROVIDERS

Sometimes farmers' groups and associations develop further than just linking smallholders and service providers, and become service providers in their own right. Sometimes organisations are set up specifically to provide commercial services to their members; these are often registered under cooperative legislation. At other times, commercial service provision is only one part of a wider function of the organisation.

Running a commercial service, and using profits from this service to fund the other activities of an association, is a possible way of achieving sustainability and therefore has great attractions. Some organisations (for instance the Likwama Farmers' Cooperative Union – see Box 7.2) have been successful in developing service provision alongside other activities. There can, however, be pitfalls:

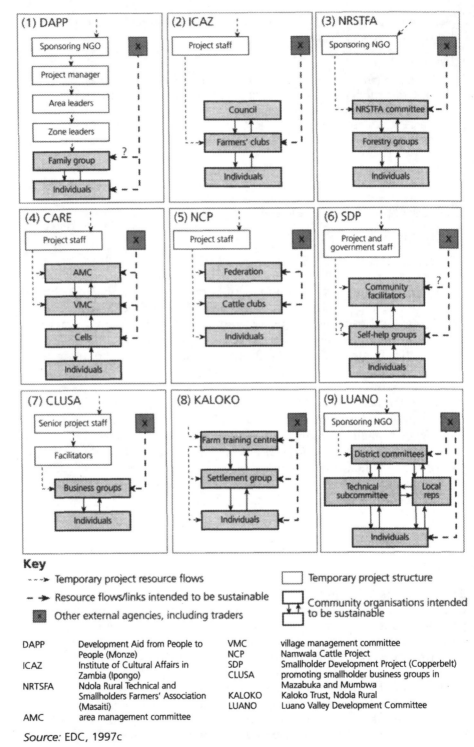

Key

····▸ Temporary project resource flows

~ ▸ Resource flows/links intended to be sustainable

◼ Other external agencies, including traders

☐ Temporary project structure

Community organisations intended to be sustainable

DAPP	Development Aid from People to People (Monze)	VMC	village management committee
		NCP	Namwala Cattle Project
ICAZ	Institute of Cultural Affairs in Zambia (Ipongo)	SDP	Smallholder Development Project (Copperbelt)
		CLUSA	promoting smallholder business groups in Mazabuka and Mumbwa
NRSTFA	Ndola Rural Technical and Smallholders Farmers' Association (Masaiti)		
		KALOKO	Kaloko Trust, Ndola Rural
		LUANO	Luano Valley Development Committee
AMC	area management committee		

Source: EDC, 1997c

Figure 7.1: *Models of NGO Interventions*

- There can be conflict of interest between business needs and the interests of members.
- Sustainability can be jeopardised by too much money being taken out of the commercial side of the business to cover other non-profit activities.
- Sometimes donor funds for capital equipment and start-up running costs create unfair competition for local commercial competitors (eg putting existing local grinding mills or input suppliers out of business).

FARMERS' UNIONS AS POLICY ADVOCATES

Throughout Southern Africa, smallholder farmers are organising into unions, usually on a national scale, to advocate policies and services that more closely meet their needs. Some of the main issues relating to farmers' unions and sustainability are:

- *Defining and representing the constituency:* smallholders cover a very wide range of household and farm types, and it is difficult for one union to represent them all – yet there is considerable pressure to present a united front for strong and unified advocacy. There is an understandable tendency for better-educated, better-resourced and male smallholders to be overrepresented in decision-making structures. Given the many immediate pressing needs, it is also unrealistic to assume that smallholder unions will automatically have an orientation in favour of longer term sustainability and the poorer smallholders. Positive actions can help to counteract, but are unlikely to eliminate, these tendencies. Examples might be: quotas (eg for women, for smallholders without cattle, for isolated districts etc.); specific committees (eg women's committees); specific training (eg for women and youth); specific actions prioritising the long term and sustainability, including joint work with specialist organisations working on these issues.
- *Strategic alliances:* most countries have powerful unions representing large-scale commercial farmers; smallholders have often formed their own union, recognising their different interests. The smallholder unions represent large numbers of people, but are often poorly resourced in comparison to the large-scale farmers' unions. The potential for collaborative work – using the organisational capacity of the large-scale union and the political mobilising power of the smallholder union – is considerable, but on some key issues the interests of smallholders and commercial farmers are clearly opposed. In KwaZulu Natal and Eastern Cape, the African Farmers' Unions have recently merged with the Commercial Farmers' Unions. In Zimbabwe and Namibia, smallholder unions have remained distinct from large-scale farmers' unions.

Box 7.2 Likwama Farmers' Cooperative Union, Namibia

The Likwama Farmers' Cooperative Union (LFCU) is an umbrella organisation serving a fluctuating paid-up membership currently numbering about 800 farmers, organised in some 25 local associations in the Caprivi region. It was first established in 1984 as a lobbying group, and is now one of the most developed organisations of its kind operating in Namibia's communal areas. The LFCU started selling agricultural inputs (seeds, fertilizer, ploughs, and other implements) to its members and began purchasing their produce in 1991. It also established a service mill for maize, millet and sorghum processing. Its local association stock agents act as marketing agents for the major livestock buyer, the private company MeatCo.

The LCFU has put considerable effort into developing its own structures. It has organised training courses in leadership skills for local association management committee members, bookkeeping for local association treasurers and managers, and business management for management structures. It also organises workshops on livestock production and marketing for its stock agents, and on crop diversification and cash crops for its crop agents. This training has involved inputs from a wide range of governmental, non-governmental, and private sector organisations; for instance, the local state veterinarian has trained stock agents and sales persons (16 local associations operate shops) to take on the selling and administering of non-prescription veterinary drugs.

As a pioneer in its field, the LFCU had to expend considerable effort in advocacy work at the national and local level towards creating an enabling environment for its own and other organisations' development. For instance, it led the campaign (through the commissioning of independent consultants' reports) to halt the sale of seeds and fertilizers and to provide subsidised tractor hire services through the government's extension service (see Box 10.3). While the extension service had previously lobbied for the same ends, in order that it could concentrate its resources on information and advisory type work, its efforts came to nothing. That these government services provided unfair competition with farmers' organisations, and thus prevented them from delivering these services themselves, was the crucial argument needed to sway the cabinet.

Likwama's experience provides a number of pointers for other farmers' organisations in the region:

- The importance of good leadership and committed membership, based on effective services, appears to be of paramount importance for success.
- It has strong involvement in advocacy at the national and regional levels (it was one of the prime movers in the establishment of the Namibian National Farmers' Union).
- It has tried hard to steer clear of party political involvement.
- It has been able to exploit windows of opportunity when its own interests have coincided with those of government (as in the cases of the reform of the government's extension services and the development of the government's cooperative support services).
- It has devoted much effort to developing its membership's organisational skills and its management structures (for instance, it ensures the representation of women in decision-making structures).
- It provides a wide range of services which are financially attractive to its members, and which earn sufficient money to support staff remuneration.
- It provides extension advice to farmers.

Source: EDC, 1997d

- *Maintaining healthy roots:* most smallholder unions have experienced a degree of top-heavy development; sometimes, as in the case of Namibia, this has almost destroyed them. Therefore, engagement in national-level politics, access to funds independent of member contributions, and the development of a cadre of paid officers have all tended to create a gulf between farmers and their national representatives, and there is a natural tendency for the apex structure, based in the capital, to become distant from the roots. Continual positive action is needed to counteract this. Examples of action include: continuing training for local association members; financial systems to be as decentralised and as bottom up as possible, with local associations keeping a proportion of membership contributions for their own needs; fixed terms of office for the main posts, to enable new blood to enter; effective separation of power between paid officers and committee members (farmers).
- *Funding:* the three common sources of funding are member contributions, donor funds and marketing levies. The balance of funds available at national, compared with local-association, level continues to be a contested point in many unions. Care is needed by donors not to exacerbate existing centralising tendencies by funding the centre at the expense of the periphery – it may be beneficial to fund the periphery, and therefore build capacity from the bottom up. Marketing levies tend to be collected from all smallholder sales, not just from members; while there is some justification for this, because of the 'freeloader' issue (with all smallholders, not just members, benefiting from activities of the union), the levies may become contentious if unions are considered unrepresentative or ineffective. Liberalised markets may also make the collection of such levies more difficult, as the number of marketing channels proliferate, and thus threaten the sustainability of some unions.

Unions do not necessarily confine themselves to an advocacy role; for instance, the Zimbabwe Farmers' Union combines advocacy with direct service provision and linking members to external service providers. It also illustrates many of the issues mentioned here (see Box 7.3).

CONCLUSION AND RECOMMENDATIONS

Farmers' groups can concentrate on a number of different roles – undertaking specific tasks, providing linkage to service providers, providing services themselves, or advocacy. There are many different models of group organisation since these depend on local circumstances. Organisational development and capacity building within farmers' organisations is crucial – structures need to develop in tandem with the communities and are likely to evolve over time. The sustainability will depend on whether the benefits of organisation are perceived to compensate for the costs; this will often hinge on whether the external environment is supportive or not.

Box 7.3 The Zimbabwe Farmers' Union (ZFU)

The ZFU was formed in 1991 from the merger of the Zimbabwe National Farmers' Union (ZNFU), representing the small-scale black commercial farmers, and the National Farmers' Association of Zimbabwe (which grew out of the National Association of Master Farmers' Clubs), representing communal area farmers. There are also two other unions representing large-scale farmers. The different unions have served on various policy-making structures, such as the boards of parastatals and the Agricultural Research Council, and contributed to favourable policies after independence for their constituents – such as new roads to communal areas, marketing depots in communal areas, favourable prices and credit for smallholders.

Structures

There are a number of different levels of organisation in the ZFU, with some diversity inherited from the base structures of the two parent unions:

- clubs: over 6,000 with typical membership of 15 to 40 farmers;
- commodity groups: often involving slightly larger-scale farmers;
- area (ward): area councils representing clubs and commodity groups;
- district: district ZFU council; the ZFU have been trying to strengthen their organisational capacity at this level – there are now ZFU offices with paid official(s) in 40 of the 57 districts;
- province: a provincial committee, commodity committees and staffed office;
- national: annual congress, national council, subcommittees, commodity committees and paid staff.

Running through the structure is a generalised hierarchy and specific commodity interest groups, representing farmers commercialising those commodities. There have been moves to decentralise decision-making since 1991. A change came in 1997, with the appointment of a new president who is a genuine communal farmer and who is reputed to be more committed to decentralisation.

Membership

The ZFU has a very diverse membership that can be broken down in a number of different ways:

(1) **Farmer type:** communal area, resettlement area, small-scale commercial, large-scale commercial (indigenous) and peri-urban plot holders. Even within these categories, there is enormous diversity, particularly in the CAs, where farmers range from really small-scale subsistence households, to those with considerable resources and producing on a significant scale for the market.
(2) **Membership type:** the ZFU divides members into active (those who have paid fees) and inactive – but, for the purpose of this analysis, further subdivision is helpful:
     - Active: have paid a membership fee to be part of the ZFU and/or are involved in various ZFU committees or activities. These members have the right to vote.

- Nominal: have paid a membership fee purely to make use of sales tax advantages*, and have no other involvement. These members have the right to vote.
- Inactive: affiliated to ZFU clubs but have not paid fees. They have no right to vote.
- All other smallholders: these are not technically members, but some feel the ZFU represents their interests; at national level ZFU is often considered to represent smallholders in general – and the levy income of ZFU comes from all smallholder sales, so they all contribute indirectly to ZFU. In addition, many of the policy changes fought for by the ZFU benefit all farmers, not just ZFU members.

The ZFU claimed an (active and inactive) membership of 700,000 (nearly 60 per cent of smallholders) in November 1997, although this seems somewhat optimistic. In 1993 and 1994 the active and inactive membership was reported as 164,000 (Arnaiz et al, 1995). In 1997 membership cost Z$20 (US$1.6) per year.

Activities

ZFU has an extraordinarily wide range of activities, many of which go beyond traditional union activities. Some of the more important are listed below:

- **Lobbying:** the ZFU does research and makes representation on a wide range of issues relating to agricultural policy and prices.
- **Research and extension policy:** this is a more specific form of lobbying in which ZFU has a place on the Agricultural Research Council and similar bodies. This is likely to become more important with changes in Ministry of Agriculture management systems, making research and extension more demand driven. In addition, ZFU has used donor funds to specifically support governmental research.
- **Extension access:** ZFU groups are active at a local level to ensure that they and their members get access to Agritex advice and support.
- **Inputs:** ZFU, particularly at district office level, is involved in bulking up input orders and negotiating volume discounts from suppliers. ZFU is also linked to the Farm & City Centre cash card scheme (see below).
- **Marketing:** organises marketing meetings at area level; supports the development of farmer marketing collaboration at the local level, putting ZFU groups in contact with potential purchasers; supports a market capacity-building venture in the Midlands. Market information is provided through a weekly radio programme (Shona and Ndebele); a monthly magazine (mainly English with some Shona and Ndebele); and an occasional marketing news sheet (currently in English).
- **Training:** this deals with organisational issues, such as leadership training and gender awareness, and technical training – for instance, through the production of technical manuals.
- **Commercial:** the ZFU has a transport fleet that it operates on a commercial basis, but with an emphasis on serving communal areas. ZFU also has share holdings in a number of organisations including Farm & City Centre.

---

* Farmers have exemption from 17.5 per cent sales tax on a range of farming inputs. Resettlement and CA farmers usually 'prove' to the shop that they are farmers by presenting a ZFU card (although being a member of ZFU is not officially a requirement, it seems to be one in practice). The whole process is a bit of a sham, but superficially beneficial to ZFU.

ZFU policies and activities are orientated towards the commercialisation of small-holder farming; these policies are very similar to the objectives of the Ministry of Agriculture.

Finances

There are three main sources of income for ZFU:

- Levies: representing nearly half of ZFU income, these come from a 1 per cent levy on communal and resettlement area sales. This has been complicated by the recent liberalisation of marketing, giving more channels and less division between communal and other areas. It is not clear if private buyers will continue to pay the levies.
- Grants: these represent approximately 40 per cent of income, mainly for specific projects supported by donors.
- Membership fees: these represent 10 to 20 per cent of income.

Issues

The ZFU is a large and complex organisation which is also dynamic, responding to a rapidly changing environment. ZFU is often maligned by NGOs as unrepresentative of poorer farmers, and while this is partly true, the reality is more complex and deserves more analysis.

(1) **Representation:** the ZFU tends to represent the less resource-poor smallholders, which is hardly surprising considering the origins of the parent unions, the specific advantages of membership for those farmers purchasing more inputs and selling more of their crops, and the nature of rural Zimbabwe, in which those chairing committees are often slightly better off, better educated and men. Many development projects show similar biases towards the educated, better off and men in representational structures. A recent study concluded that the bias towards the better off was probably overestimated by ZFU's critics (Arnaiz et al, 1995).

The problem, perhaps, is not so much that the poorest are underrepresented, but that ZFU alone is expected to do the impossible and represent such a diverse group of farmers. Therefore, the presence of ZFU (along with the large-scale unions) on, for instance, the new Agricultural Research Council, with the implication that the ZFU represents the interests of all smallholders, is insufficient. ZFU needs to be joined by other groups working with the very poor; the ZFU should see such other organisations as allies, rather than as a threat to its sovereignty.

At the same time, continual affirmative action is needed to encourage increased participation by the resource-poor in the ZFU. There is some indication that such action is being taken more seriously for gender issues, but the issue is actually broader (see below).

(2) **Orientation:** closely linked to representation is the orientation of ZFU – particularly towards smallholder commercialisation through increased use of external inputs and market integration. This reflects the aspirations of many, especially the better-resourced smallholders. The problem is not that ZFU's actions on inputs and marketing are wrong, but that they are not sufficiently balanced by an emphasis on sustainability and the needs of the resource-poor rural farmer.

(3) **Finances and membership:** this is clearly a sensitive area. With union membership there is often a free loader problem, in that some benefits achieved by the union (eg policy changes) go to all, not just to those who have paid their fees. This perhaps justifies the levy on all smallholder sales, rather than just on members sales, and the sham of needing a ZFU card to obtain sales tax exemption.

The sales tax procedure has the advantage of encouraging membership – and once they have joined, perhaps farmers are more likely to make demands and become active, even if this is not their original reason for joining. The downside of people joining in order to avoid sales tax is that it takes the pressure off the ZFU to provide more genuine benefits for members, and it is difficult to judge the real level of support for the ZFU in communal areas.

(4) **Conflicts of interest:** although it is not suggested here that the ZFU has succumbed to conflicts of interest, there is potential for confrontation, particularly in relation to increased use of external inputs. The ZFU clearly benefits from greater use of inputs by smallholders since it encourages membership to avoid sales tax and because of profits from the use of Farm & City cash cards (ZFU gets a small percentage of the turnover from the use of these cards). There is no similar incentive to encourage low-input approaches.

(5) **Competing unions:** the need for a unified voice for indigenous farmers has resulted in pressure against free competition between unions. Not only did government encourage the merger of ZNFU and the National Farmers' Association, but government also actively suppressed local initiatives of farmer organisation, independent of ZFU (EDC, 1998a). For longer term sustainability, more freedom to organise will be needed, enabling farmers to choose which apex structure best meets their needs.

Conclusion

The ZFU provides necessary organisational and practical support to many farmers and it is important that it continues to develop as a strong and independent organisation. Doubts about whether it is wholly representative are founded to a degree on unrealistic expectations; ZFU should not be expected to act as the sole voice of smallholders. More emphasis by ZFU on long term sustainable farming and the needs of the poorer smallholders could lead to a more balanced orientation and facilitate more productive partnerships with NGOs at a local level.

Source: EDC, 1998a

There are some advantages for farmers' organisations to operate as service providers, not least because the organisation should be well aware of the type of services required by their members and the provision of services can provide a way of funding long term sustainability. However, to be successful, this calls for a hard-headed approach and the ability to overcome any conflicts of interest. Lobbying is needed to improve policies and services for smallholders, and smallholder unions have an important role to play in this. However, it is unreasonable to expect a smallholder union, representing a diverse membership, to necessarily give priority to the issue of long term sustainability and the concerns of the poorest smallholders. Positive action and alliances with other organisations are required to achieve this.

Chapter 8

Community-Based Natural Resource Management

INTRODUCTION

Community-based natural resource management (CBNRM) has caught the imagination of policy-makers throughout the region. Building largely on the experience of the pioneering schemes in Zimbabwe (CAMPFIRE) and Zambia (ADMADE), CBNRM is given prominence in national environmental action plans and nature conservation strategies in every country. Every ministry concerned with the environment has some CBNRM projects underway or planned. The concept is radical – instead of top-down, government-imposed controls to protect the environment, which were often not supported by local communities and often impossible to enforce, it is argued that the community, which has a vested interest (or can be given a vested interest) in sustaining the local environment, is given the responsibility to do it in its own way. There are some fundamental preconditions:

- It is necessary for a (relatively) discrete community to exist with a defined relation to particular natural resources.
- That the community's diverse interests (of rich/poor, men/women, grazers/hunters/crop farmers, etc) must be sufficiently compatible.
- That the benefits of collective action must outweigh the costs.
- That the community must have the capacity to manage the scheme.
- That the community must have the capacity to enforce rules among itself and on outsiders.
- That the community interest in the environment must be sufficiently compatible with national (and often international) interests.

WORKING WITH TRADITIONAL LEADERS

CBNRM fits into a wider picture and history of landuse planning and management. Traditional landuse planning was, and still is, organised by local communities, chiefs and spiritual leaders – for instance, deciding who can settle where, which land can be used for crops and grazing, and which is to be preserved as sacred. There is now greater appreciation of the importance of, and willingness to work with, traditional institutions. One attempt to do this more systematically has been by the Association of Zimbabwean Traditional Environmental Conservationists (AZTREC) (see Box 8.1).

CURRENT EXPERIENCE IN CBNRM

The experience of CAMPFIRE (Communal Areas Management Programme for Indigenous Resources) in Zimbabwe is well known and has been well documented in many reports, particularly from the Centre of Applied Social Studies (CASS). Despite many successes, it should, however, be noted that CAMPFIRE is:

- Still largely wildlife (tourism and tourist hunting) based: experience from wildlife-poor areas, with management of resources such as grazing, firewood and thatching grass, is less complete.
- Despite some well-publicised examples of excellent community returns, the amounts of money involved are, in all but a small number of wards, insufficient either to have a significant impact on individual household income or to compensate for the damage done by wildlife (Bond, 1993; Thomas, 1995b).
- Despite a large amount of resources devoted to community capacity-building, capacity remains variable and in some communities insufficient (Conyers pers comm).
- Many CAMPFIRE communities still retain little effective control over their natural resources (Thomas, 1995a).

The purpose of these observations is not to reject CBNRM, but to avoid adopting the approach as a magic bullet.

In Botswana, one of the main experiences with CBNRM is the Chobe Enclave Project, which, although fairly new, is reported as being successful (Modise, 1996). In Malawi, the National Environmental Action Plan is being complemented by an environment support programme which has created a mobilisation and micro projects fund. NGOs will be involved in training villagers, who will then form village natural resource committees; so far there has been little practical experience, but there are some interesting results from community consultations (EDC, 1997b). The blueprint approach to organisation, however, raises suspicions that this may be an example of top down defined participation, rather than organisation devel-

---

Box 8.1 Association of Zimbabwean Traditional Environmental Conservationists (AZTREC)

AZTREC's vision is to see that 'indigenous knowledge systems serve as a guiding torch towards sustainable development'. AZTREC works in the Masvingo area of Zimbabwe and was formed by people who had been active in the liberation struggle. During the liberation war, the combatants often worked closely with the spirit mediums and traditional leaders. Both are traditional custodians of the environment, with the spirit mediums providing a communication channel between the present population and the ancestral spirits, and the traditional leaders defining how natural resources should be used. Much of the current environmental degradation is considered to be caused by a lack of respect for traditional environmental management practices.

AZTREC started work in 1985 and is involved in a number of programmes including:

- research and promotion of indigenous technical knowledge;
- promotion of nurseries and distribution of tree seedlings, including indigenous species (distributing over one million seedlings to 21,000 people);
- demonstrations of natural farming;
- reviving the tradition of *Zunde Ramambo* or granary of the chief, in which land is cultivated communally for the chief and the food store created can be used by the needy in times of drought;
- water harvesting techniques;
- water source and catchment management;
- wetland conservation.

Although many of the activities of AZTREC are not very different from those performed by other NGOs working on the local environment, the origins and explicit linkages with traditional structures, with a board largely made up of spirit mediums and traditional leaders, is somewhat unique. This gives AZTREC particular advantages in working with local communities and is probably an approach which could be more widely adopted in the region. AZTREC's traditional linkages have not meant that it has ignored youth and gender issues; it has been active in mobilising young people in environmental conservation, particularly in its work with schools. Women form the majority of most AZTREC groups and there were 24 women in the 27-member AZTREC council in 1995 (AZTREC, 1995; 1996; Mema & Murombezi, 1996).

AZTREC is currently negotiating with a range of other organisations in the Masvingo area (funded by Oxfam–UK) to develop an integrated food security programme. This will mean that the various skills of the different organisations will be used in a more systematic way; AZTREC will therefore bring its specific orientation to a wider programme, while in turn benefiting from specific technical skills being developed by other collaborators.

Source: EDC, 1998a

---

oping out of local experience and diversity – this is not uncommon when large organisations such as the World Bank or national government try to institutionalise participation.

Box 8.2 Development of CBNRM Policy in Namibia

Until 1967, all wildlife in Namibia was the property of the state, and wildlife numbers outside the national parks were in decline. In 1967, the ownership of wildlife on the freehold farms was transferred to the landowners and sustainable use of this wildlife was permitted. These farmers began profiting from wildlife, and wildlife numbers generally started to increase.

After independence, the Ministry of Environment and Tourism (MET) moved to extend the rights over wildlife enjoyed by private farmers to communities in the communal areas, using a two-fold strategy:

- local pilot projects enabling communities to benefit from and manage wildlife;
- development of policy and the revision of legislation.

Legislation was enacted in 1996, which enables a community to register an area of land as a conservancy: the community has the right to benefit from wildlife through tourism, sale of live game, trophy hunting and sustainable harvesting quotas agreed with MET. To be registered by MET, a conservancy must have:

- defined boundaries (involving dialogue with neighbouring communities, regional councils, etc);
- defined membership (none to be excluded on grounds of ethnicity or gender);
- a representative committee (the community is allowed to choose its own system of representation, as long as it is accepted by the membership);
- a legal constitution (including a set of rules establishing how the conservancy will operate);
- a plan for the equitable distribution of the proceeds and a sound accounting system;
- satisfied the MET that it is able to sustainably manage the wildlife (a complete management plan is not needed at first).

In order to build community capacity to manage a conservancy, a process of community development is usually needed. This process often involves a specialist NGO working alongside the community.

Source: EDC, 1997d

In Zambia, the CBNRM programme ADMADE seems to have been unable to stop poaching; however, that is perhaps an unreasonable expectation, as neither has top-down enforcement. The process by which Namibia has developed its CBNRM policy is given in Box 8.2, and an example is given in Box 8.3.On a completely different scale, Box 8.4 looks at the way a Zambian village tree nursery group is combating local deforestation, while benefiting its own funds.

The project in Masaiti (see Box 8.4) in Zambia is interesting on a number of counts: although progress towards some of the physical outputs of the project has been slow, environmental activist behaviour has developed relatively unplanned in at least one nursery group, and conditions seem to have been developed which provide incentives to enforce environmental protection, even if on a very limited scale.

BOX 8.3 SALAMBALA CONSERVANCY, CAPRIVI, NAMIBIA

The Caprivi region is rich in wildlife, yet – with the exception of fish – the local population derive little benefit from it. The main sources of livelihood are crop and cattle production, and wild animals cause considerable damage; the communities feel that the government does little to protect them from these animals, while preventing the community from killing them.

In 1990, an NGO, Integrated Rural Development and Nature Conservation (IRDNC), started working with a number of rural communities in Caprivi. Initially, the response of communities was hostile since they had had negative experiences with government wildlife protection programmes. IRDNC began working with the local communities and traditional leadership on a game guard programme to reduce the damage caused by 'problem' animals to the communities; guards were chosen by communities and paid by IRDNC. As the communities have experienced the benefits of the programme, trust has grown; currently a number of communities are working with IRDNC on developing conservancies in their areas.

The Salambala community is an example in which the population of about 6000, in cooperation with the traditional leadership and IRDNC, has developed a plan for a conservancy on the 80,000 hectares in which they live and farm. This will involve establishing a core wildlife area of 14,000 hectares in which no agriculture or grazing will be allowed. The community is negotiating a joint venture agreement with a tourist company in which:

- The company will develop a lodge and gamebird shooting enterprise in the core area.
- The company will pay a concession fee of N$30,000 (US$6800) and a percentage of turnover to the community.
- A levy will be paid to the community on gamebirds killed.
- Local people will be employed and trained by the company whenever possible.
- After ten years the fixed assets will be transferred to the community.

The community is developing a trust to look after the money from the conservancy, which in addition to members of the community will include a representative of the traditional authority, a church leader and a representative from an outside organisation. A separate management committee will organise the running of the conservancy and the distribution of benefits to the community – receiving money from the trust to do this.

Projects like this are, however, rarely conflict free. In mid 1997 disagreements over the approval of the use of the land in Salambala surfaced unexpectedly and were putting the whole scheme in jeopardy.

Source: EDC, 1997d

COMMUNITY-BASED GRAZING MANAGEMENT

Community-based grazing management is a particularly important type of CBNRM. Smallholder grazing land is still communally used in Southern Africa. Although there has been considerable private fencing in some areas, the costs of individual fencing and water provision in the dry, and there-

---

BOX 8.4 FOREST GROUPS IN MASAITI, ZAMBIA

Although Zambia is comparatively well endowed with forest resources in relation to its rural population, a high level of urban demand for charcoal has resulted in localised deforestation in more accessible areas. Until recent times, Masaiti district (formerly part of Ndola Rural) was covered by dense forest, and the Lamba-speaking inhabitants were predominantly hunter–gatherers, rather than farmers. However, this is now changing rapidly under the combined pressure of agricultural settlement and increased demand for charcoal from Lusaka and nearby Copperbelt towns. Charcoal is produced by long-standing residents, newly arrived settlers clearing bush for the first time, and by non-resident charcoal burners and traders.

Responsibility for forest outside gazetted national reserves is officially shared by the district forestry officer (DFO) and local chiefs, with those wishing to sell charcoal required to secure permission from both. As demand has increased, so the system for managing local resources has weakened, both by the DFO's lack of resources and by rivalry (in the form of overlapping claims on people and land) between chiefs. Awareness of the need for improved forest management nevertheless appears to be quite strong. For example, a participatory evaluation of the EU-sponsored smallholder development project in the area in 1994 identified reforestation as an important priority, even though it was not a formal project component (Nkomesha et al, 1995: 19; Harvest Help, 1994a: 9).

It is in this context that the Ndola Rural Technical and Smallholder Farmers' Association (NRTSFA) was set up in 1993. The main inspiration for its establishment came from the DFO and five other government 'technicians' based in Masaiti, seeking ways to mobilise funds to supplement the limited recurrent resources available to them through their respective departments. The executive committee also included five farmer representatives, and in 1994 the association had 235 members (Harvest Help, 1994a). NRTSFA's stated objectives include promotion of sustainable landuse, diversification of livelihoods (particularly into fruit production), and promotion of agroforestry. At the end of 1994, NRTSFA embarked on an agroforestry and natural resource management programme with support from the NGO Harvest Help, and a budget of 36 million Kwacha over three years (approximately US$48,000).

Progress towards physical targets (nursery group formation, seedling production, well construction) during the first two years of the project has been very limited, although this can be attributed partially (perhaps largely) to delays in disbursing adequate funds. More positively, at least one tree nursery group (Zakeyo) appeared to be redefining itself as a forestry protection committee, with a view to playing a more active role in enforcing controls on charcoal burning. The group had already helped to apprehend one unlicensed charcoal burner, and was also trying to establish the principle that anyone clearing land in its ward should offset the trees they cut by buying and planting seedlings from its nursery – at a rate of 100 seedlings per hectare cleared. The group was therefore trying to tie up its interests in selling trees with locally defined rules on the use of tree resources. In this way its vested interests were helping to provide the incentive for natural resource protection.

Source: EDC, 1997c

---

fore extensive, grazing regimes is prohibitively expensive for individual resource-poor households.

Open access grazing has been extensively criticised as being uncontrolled, leading to overstocking, low productivity and environmental destruction. Concerns about range degradation and data showing low productivity from communal systems, such as those in Table 8.1 from Botswana, have been used to shift policy towards private leasehold tenure.

Table 8.1: *Productivity in Communal and Leasehold Areas in Botswana*

	Communal Area	Fenced Area
Calving rate	50%	60%
Annual mortality	12%	5%
Annual offtake rate	8%	17%

Source: Government of Botswana, 1996

While pertinent, these sorts of figures only show part of the picture:

- Productivity improvements are not necessarily due to fencing – those that can afford fencing also can afford other inputs to improve livestock productivity and reduce mortality.
- The offtake rates do not include many of the other valid contributions which livestock make to the rural (and therefore national) economy, such as draught power, milk, savings and insurance.
- No consideration is given to the overall productivity of the land – for instance from veld products.
- There is no consideration of costs (financial and economic) of the different systems, nor their employment or self-employment potential. These are all needed in order to make a realistic comparison of communal grazing with fenced grazing.
- There needs to be greater clarification over the type of communal grazing since there is often confusion between open access and common property systems (see Table 8.2).

Table 8.2: *Types of Access to Grazing and Other Natural Resources*

Type of Ownership/Access	Used by	Degree of Management
Open access	Anyone	Low or non existent
Common property	Defined community	Variable – depends on traditions, knowledge, organisational capacity and incentives
Private	Individual, household or syndicate	Variable – depends on knowledge, capacity and incentives
State	Variable	Variable – depends on policy, knowledge and enforcement capacity

The table indicates that common property is not inherently poorly managed; much depends on community traditions, capacity, knowledge and incentives. Many common property regimes have been undermined in recent years, particularly since traditional authorities have lost power and have tended to move from a common property regime towards open access; the degree of management has consequently fallen.

There are many past and current attempts to improve or make communal grazing more sustainable:

* *Enforced destocking/stock limits* – based on the LSU calculation (see Chapter 5): the LSU quota is then sometimes used as a top-down control mechanism to prevent 'irresponsible' farmers from seeking short term profits from overgrazing. The top-down approach to limiting stock numbers in communal areas has a long history of failure in Southern Africa; recent research and experience have raised important questions about the applicability of this approach. Despite being partly discredited by experience, this approach continues to be used throughout the region.

* *Rotational grazing/paddock systems:* these have been tried throughout the region with limited success. There has been considerable experience of grazing schemes in Zimbabwe over a number of years. These have generally involved a division of land into grazing, arable and residential blocks, and fencing of the grazing area, which is usually divided into a number of paddocks. Despite the objectives of project donors often being 'better' grazing management and environmental conservation, there has been little evidence of this (see Boxes 8.5 and 8.7) and it is important to consider the following. Communities have usually supported grazing schemes as a way of reducing herding labour (particularly important with children going to school) and reducing cattle straying onto fields in the growing season. Management of rotational grazing has either been lax or organised on a blueprint approach, defined by agricultural extension, (typically two weeks' grazing, two weeks' rest), rather than active management depending on the condition of the grass. There is little evidence of communities limiting stock numbers as a result of instituting a grazing scheme, although this is currently confused by high mortality during recent droughts, leaving stock numbers in many areas still below the optimum. Although grazing schemes have brought some benefits, they have often been expensive, in relation to the benefits achieved. A modification of this system, with movable electric fences, is being tried by the SARDEP project in Namibia (EDC, 1997d); however, it is too early to judge its impact.

* *Community managed tracking:* livestock numbers are expected to rise and fall according to grazing availability, and livestock will move over considerable distances in search of fodder and water, at different times of the year and in different years. Such schemes are usually built on traditional rules and practices. There is a need to develop

---

BOX 8.5 MHONDORO GRAZING SCHEMES, ZIMBABWE

The first grazing scheme in the Mhondoro area was started in village development committee (VIDCO) 1 in 1984. The councillor, who is now Chief Rwizi, was instrumental in persuading people of the benefits of grazing schemes. Before this, the local population was suspicious, believing grazing schemes were being promoted by the government in order to restrict the number of their cattle.

Each member of the VIDCO contributed one bucket of maize, amounting to a total of 30 to 40 bags which they sold to the grain marketing board (GMB). However Agritex and the District Development Fund (DDF) produced the fencing material without using the money collected; this instead was used to organise a party when the fencing was completed. Agritex mapped and planned the area and the community installed the fencing. Six paddocks were made and these were divided: two to each *kraalhead*. Grazing is organised by each *kraalhead* with two weeks in each paddock. Those people who had previously been living inside the designated grazing area were either expected to move or to fence their homesteads. All VIDCO members helped with the fencing, whether they owned cattle or not, since they all perceived themselves as potential beneficiaries. The fences continue to be repaired by all VIDCO members, whether cattle owners or not.

VIDCO 3 followed with a similar grazing scheme, and they were followed by VIDCO 2 in 1990. However, by then funding was more difficult to obtain, so there was only material for four paddocks. Finally, in 1995, VIDCO 4 started a grazing scheme – but there were only resources for the perimeter fence, and so no paddocks have been made.

The principal advantage of the grazing schemes, in the opinion of the participants, is that less time is needed for herding, giving more time for adults to cultivate and young people to go to school. There has been a degree of protection of trees in the grazing area, probably because of the prohibition on clearing for cultivation; bush fires also tend to be controlled in the paddocked grazing area. However, neither of these were cited by participants as benefits of the scheme.

Agritex set a maximum stocking level for each grazing scheme at the initiation; it seems that these levels were generally exceeded before the 1992 drought, but are probably not currently being exceeded. Although members do not appear to question the limits set by Agritex, there is no enthusiasm for trying to stick to them. There was little evidence that the development of the grazing scheme had encouraged self-reliance. Members made insignificant contributions in cash terms to the scheme, although they provided labour when required. Fences were maintained using poles and droppers left over from the original grant – when these are finished the chairman plans 'to go back to the DDF to ask for more'. The members have a number of plans for improving the scheme, including enriching the pasture with more palatable grass, introducing improved bulls and opening new water points. However, in each case, the strategy is to wait for the DDF to provide the funds.

Source: EDC, 1998a

---

experience with this within a more supportive policy environment, in which communities are legally empowered and given the capacity to manage their own grazing.

There are various possibilities for locally negotiated solutions to specific grazing crises experienced by communities. Sometimes these do not offer a comprehensive solution but do relieve a particular bottleneck; an example is CAP Farm Trust in South Africa, where sustainably community-managed dry-season grazing is meeting a key need, even though the wider area is under an open access arrangement, with little evidence of active management (see Box 8.6).

COMMUNITY LANDUSE PLANNING

In all countries in Southern Africa there have been a number of top-down landuse planning interventions (including, in colonial times, the allocation of much of the best land to settler farmers). There have been various programmes aimed at planning landuse to conserve natural resources or facilitate service provision. An example was the policy of 'villagisation' in Zimbabwe after independence, which had the objective of bringing households closer together for services, and of instituting what was considered a more sustainable division of land between residential, crops and grazing (see Box 8.7). The overall impact was quite limited and the programme seems to have faded away. Interestingly, a form of landuse planning seems to be returning in Zimbabwe, and similar initiatives seem to be being planned, or are underway, in other countries in the region; therefore, lessons from the past need to be revisited.

Heightened environmental awareness means that community landuse planning is returning to fashion in Zimbabwe with the new name of District Environmental Action Planning (DEAP) (see Box 8.8). This started in 1995 and is designed as a bottom-up response to the National Environment Action Plan (NEAP), the National Conservation Strategy and Agenda 21. The DEAP process is executed by the Department of Natural Resources through task forces convened at district and ward level, and with support from UNDP, IUCN and IDRC. DEAP originally started as a pilot programme in selected wards in eight districts; there are now plans to expand it to a national programme, which will involve decentralising much of the currently centralised support functions to provincial level.

DEAP has some similarities with earlier community land use planning programmes, but also some important differences. DEAP aims to be more rooted in a participatory approach, incorporating many of the techniques and methods developed from RRA and PRA in the last ten years. DEAP uses a number of specific tools to enable a community to look at sustainability issues – in particular, it uses the analogy of an egg (the 'egg of sustainability') to explore the relationship between the human condition (the yolk) in a community and the ecosystem (the white), demonstrating that both need to be in a vital condition in order for the whole to be healthy. DEAP has developed a package of training materials and literature to back up its work in the field.

BOX 8.6 RESOURCE MANAGEMENT AND CONFLICT AT MDUKATSHANI, SOUTH AFRICA

Mdukatshani farm, managed by the CAP Farm Trust, is situated in the Tugela Valley, an area in which resource shortage, deprivation, conflict and violence are endemic. As a result, negotiation over resource use has been a key function of the project staff and one that has been fraught with difficulty.

Mdukatshani has evolved a unique and highly successful approach to resource sharing with the wider community. The provision of resources to the residents of neighbouring communities has centred around access to winter grazing, thatching grass, firewood and medicinal plants. In the tribal area, livestock deplete the grass cover and browse heavily on the trees and shrubs throughout the summer. Without any winter reserve, livestock was often in poor condition by the arrival of the spring rains.

Most of the Mdukatshani rangeland lies on a plateau above the Tugela Valley. In conventional terms, it has been assessed as requiring between four and eight hectares per large stock units (LSUs): the farm is capable of carrying between 312 and 625 head of cattle. In fact, 15 tenant households run 180 cattle on a 825 hectare portion of the farm all year round, and Mdukatshani provides its neighbours with access to the remaining 1675 hectares of its grazing lands each winter, in exchange for cooperation in regulating and maintaining the use of the resource.

The use of the grazing resource is negotiated with the user groups, under the authority of the chief Indunas (tribal prime ministers) of the two neighbouring chiefs. These chief Indunas have also been trustees of the overall project since its inception. Livestock owners utilising the grazing are organised into seven user groups, each of which has access to a defined, camped area of the grazing reserve. In the autumn, they organise work teams of ten people per day to repair the fences and burn fire breaks. Once this had been completed the cattle are herded into the grazing reserve, where they remain until the spring rains have come (usually from May until October). Goats also browse the area and play a significant role in controlling the encroachment of thorn scrub, as do duiker and other antelope endemic to the area.

The Mdukatshani veld is well balanced in terms of tree, shrub and grass ratios. By the start of the winter grazing season, it is richly grassed with palatable species, whereas the grass cover in the adjoining open-access veld is poor and visibly less diverse. Heavy, non-selective grazing by the large herds of cattle utilise the resource very effectively by the end of the winter, so that both palatable and unpalatable species are prevented from becoming moribund. The cycle of very heavy grazing in the winter, followed by complete rest in the summer, has had the effect of promoting the growth of palatable and nutritious species.

Effective regulation of this system is vital. The reality of the Tugela Valley is that regulation by government officials has not been a viable option and the Soil Conservation Act is not applied. The law, even when backed up by armed police, has only a very limited ability to regulate people's economic utilisation of the natural resource base. Regulation must therefore be based upon a finely tuned mix of enlightened self-interest and authority. In this case study, an active programme of community involvement has promoted awareness among the resource users of the benefits of cooperation within the resource-sharing schemes. Awareness alone has, however, proved insufficient, when authority has broken down. The active involvement of the tribal authority and the NGO in enforcing the rules has been essential.

In Mdukatshani, authority is only respected in this manner when there is general consensus that it will contribute to people's well-being.

Lessons learned

- Effective conservation needs to combine technical parameters with local social and political possibilities.
- Local arrangements, such as grazing management, need to be negotiated at a local level so that the stakeholders understand the system and have a vested interest in making it work.
- Regulation must be based on a finely tuned mix of enlightened self-interest and authority.
- Regulation by stocking rate, based on static estimates of LSUs, are not particularly helpful in highly variable environments. Heavy grazing, but for a limited time, can help to maintain palatability and forage diversity.

Source: EDC, 1998b

BOX 8.7 CHIWESHE DEVELOPMENT SCHEME, ZIMBABWE

The Chiweshe Development Scheme is situated in ward 12 of Buhera district; it involves nine *kraalheads* comprising about 600 households and covers 8835 hectares of residential, arable and grazing area in natural region IV. The scheme started in the context of the government's 'villagisation' programme in the mid 1980s, in which landuse at a local level was to be replanned, including the concentration of scattered residential and arable areas, in order to make it easier to provide services and to create a more sustainable and efficient landuse pattern. The scheme was sponsored by the government, who provided technical inputs, the Lutheran World Federation (LWF), who managed and coordinated the scheme, and the European Union who provided the bulk of the finance.

The scheme began in 1984 with the sponsors looking for an appropriate site for a 'villagisation' pilot programme; interestingly, about ten village headmen rejected proposals for a pilot scheme before the Chiweshe community were identified. The Chiweshe community did not particularly want a 'villagisation' or grazing scheme – they wanted a dip tank – but in the end a scheme was negotiated which the villagers felt would be of benefit to their cattle, and which included a dip tank. Before the 'villagisation', people lived in scattered homesteads near the river, with homesteads, fields and grazing areas mixed. Households had very variable cultivated areas, ranging from two to ten hectares; it was also apparent that there was much underutilised grazing around the fields.

The replanning scheme involved dividing up the land into grazing, cropland and residential. Originally the planners allocated the residential land at the foot of some hills; but the community rejected this since there was no water there, and negotiated a different area for housing. Cropland was allocated as 'two hectares per wife'; those richer men who had access to larger areas have tended to retain access to more land because they have more wives. A certain amount of surplus land was also allocated for future households. In addition, many households, unknown to the planners, managed to subvert the system by registering young girls for land. Grazing land is fenced and divided into 18 paddocks – each *kraalhead* having access to two paddocks.

In order to obtain this redivision of land into residential, grazing and arable, a total of 336 households had to move, with only 200 being able to remain. No compensation was paid, although the law has since changed and compensation might be required in a similar scheme today. Because people were relocated away from the river, water for domestic consumption was needed, and the project therefore put in a water supply. Since people were living closer together, pit latrines were considered necessary, and a clinic was also constructed.

There is considerable local ownership of the scheme. Only the Chiweshe community are allowed to gather firewood, cut poles or graze animals in the paddocked areas; they do have trouble with neighbouring communities who poach grazing – and efforts are made by the Chiweshe community to prevent this. Efforts are also made to control bush fires, more because of possible damage to the fencing than to damage to grazing since it is considered that the grass soon grows back. For some time after the paddocks were constructed, the leaders used to come to LWF to report problems such as fencing being stolen by a neighbouring village. LWF made it clear that this was the community's problem, and the community now deal with such problems directly without recourse to LWF. The fencing seems to have been maintained adequately by the community and the paddocks are used on a reasonably disciplined two-week rotation. The community has, however, failed to make sufficient contributions to maintain the water system – households were supposed to contribute to buy diesel for the pump but often fail to do so, meaning that water must be collected from the river. The community is now asking for a more sustainable well and hand pump system, rather than a supply reliant on diesel pumping.

Ten years on, all the inhabitants who were interviewed considered the scheme to bring benefits. However, many of the key benefits were not the ones for which the scheme was primarily designed – more sustainable management of natural resources. The main impacts cited by the participants were:

- The fenced paddocks reduced time and labour for herding, leaving more time to work in the fields. Even those who do not own cattle (about 30 to 40 per cent) consider the paddocks an advantage since they need to borrow cattle for ploughing and are also aspiring to own cattle.
- The developments brought in along with the landuse planning were well appreciated (eg a road to the scheme, clinic, water supply, etc).
- People enjoyed living closer to their neighbours (the ability to borrow from each other was mentioned).
- Less erosion was apparent partly because most fields are better sited, but also due to peer group pressure – when all the fields are contiguous – to undertake contour ridging. However, there was now a problem of falling soil fertility caused by continuous cropping of the two hectares belonging to each household, with no space for fallows or moving to other areas.
- The paddocked areas were considered to have better grazing than the unfenced area belonging to neighbouring villages, which was translated into fatter cattle and better prices when cattle were sold.
- The fenced area is more wooded than the surrounding unfenced communal grazing areas of neighbouring villages – this seems to be mainly due to the prevention of clearing for fields within the paddocks and possibly due to the prevention of fires, to preserve the fencing, rather than any other conscious management of the wood resource.

There are no current restrictions on the numbers of livestock allowed within the scheme. Before the 1992 drought it seems that the numbers were above the Agritex recommended maximum, with again no effective measures taken to limit numbers. There was considerable cattle mortality during the 1992 drought, although opinions differed on whether this was the same or less than that of the neighbouring unfenced areas. It seems that currently, with about 900 cattle on 8800 hectares, there is sufficient grass, and the cattle owners are appreciating the good condition of the cattle. The scheme's chairman claims that if cattle numbers rise sufficiently to cause overgrazing, people will voluntarily sell some cattle. It seems unlikely that this would actually happen in practice.

There is also little indication of active management of the grazing; it seems that a two-week rotational grazing cycle – as advised by Agritex at the beginning of the scheme – is used between November and April when crops are in the fields, with open grazing for the rest of the year. Therefore, there does not seem to be an attempt to change the cycle according to the rate of grass growth, or to set aside some grazing for the dry season.

The development scheme seems to have engendered an identity and an interest in development within the community. In 1997, there were plans to try to introduce some improved breeds of bull, to produce bigger cattle. There were also plans to develop a more sustainable water supply system. Some households, particularly poorer households without cattle, were aware of a crisis looming because of the fall in soil fertility on their crop lands, but felt relatively powerless to tackle this.

Neighbouring communities consider, in retrospect, that they missed out in not joining the scheme. Yet, they feel they would probably not join a new scheme if given the opportunity because:

- The disruption is so great and the benefits are only felt in five to ten years' time.
- When the Chiweshe scheme was implemented, most people lived in traditional houses; now many more people have brick houses with latrines, so the cost of moving would be much greater.

In addition, the coordinator felt that such a scheme was very expensive, in relation to the benefits gained, and therefore difficult to justify.

Source: EDC, 1998a

In practical terms, DEAP has been a vertical programme, introduced to districts from the national level. At the district level, there is a district strategy team which is officially a sub-committee of the district development committee; the district strategy team chooses a ward in which to initiate the DEAP process. In the chosen ward, the traditional and elected leadership undergo training and a community strategy team is formed. Various participative processes are then used to identify key issues concerning sustainability and a process is initiated in which the community, in partnership with outsiders, addresses the key issues. The community is encouraged to identify those things:

- that can be solved by members of the community themselves;

BOX 8.8 DISTRICT ENVIRONMENTAL ACTION PLANNING IN WARD 14, BUHERA DISTRICT

Buhera district in Manicaland is made up entirely of communal land. In ward 14, like much of the district, there is considerable pressure on natural resources, with shortage of arable land resulting in continuous cropping and consequent risks of fertility decline and erosion. There has been considerable pressure on grazing resources, although this is currently diminished, due to high cattle mortality in recent droughts.

The DEAP process at district level is coordinated by a district strategy team, involving a cross-section of disciplines, chaired by the principal executive officer of the district council and with implementation responsibility largely held by officers from the Department of Natural Resources (part of the Ministry of Mines, Environment and Tourism). In ward 14, the DEAP process has produced a number of initiatives:

- the delineation of a specific grazing area and the movement of some homesteads and fields from this area to a more consolidated cropping area;
- the construction of a dam (by contractors, but with some community labour assistance) both to water livestock, and for a future irrigation scheme;
- the restoration of contour bunds, which had originally been installed in the 1960s but after that were unmaintained;
- the filling in of some gullies with stones and the starting of a vetiver grass nursery to produce planting material to stabilise the gullies;
- the moving of individual stream bank vegetable gardens into communal gardens (with a shared fence, but with each individual having his or her own plot within the garden) further away from the stream but near hand pumps (stream bank cultivation is against by-laws as it is considered an erosion risk).

It is too early to evaluate the impact or sustainability of the developments started through the DEAP process in ward 14. The provision of a dam is considered by the community as a major advantage and there appears to be genuine enthusiasm to undertake soil conservation measures to prevent the dam from silting up. The recreation of the 1960s contour bunds is an interesting step, partly necessitated by the lack of levelling equipment within the local Agritex to repeg the area. Some bunds of doubtful orientation were, however, observed, perhaps indicating the need for more technical input and the training of farmers. Whether the bunds will be maintained this time remains to be seen.

The replanning of the region into grazing and homestead areas has interesting local political dimensions; it seems that some people who are living in the designated grazing area are not members of the local *kraal* structure and are reluctant to move. Local *kraalheads* hope to use the DEAP process to get the council to force them to move out of the area. While this might be legitimate, there is also the possibility of the 'environment' and 'participatory decisions' being misused to consolidate particular power structures. There is nothing new in this, but it points to the need for caution when using participatory approaches, including a depth of local understanding and a maturity of experience in the use of participatory methodologies, which need to be developed over time. This indicates a need for prudence in the overrapid expansion of the DEAP process, as well as the need to learn from other experiences in community landuse planning and community development programmes which have been implemented over the last few decades.

It remains to be seen to what extent DEAP will manage to be a locally run process, creating the enabling environment for sustainability, or whether like many other programmes it will become yet another vehicle for funding small projects – useful in themselves, but not sufficient to make the transformation to sustainability.

Source: EDC, 1998a

- that can be solved by the community, with outside assistance;
- that need outside assistance.

The community and the DEAP team then work on a combination of these.

CONCLUSION AND RECOMMENDATIONS

To be effective on a bigger scale in the region, and across a wider range of natural resources, it will be necessary to scale up CBNRM, from a project to a programme approach. This will involve:

- Legal and policy changes: which are sometimes needed in order to include a wider range of natural resources in CBNRM. Land law and CBNRM law must be compatible, giving communities tenure security over common land and the natural resources associated with the land. Work on defining legally recognised community boundaries and structures may be required. Communities need to be able to punish offenders and exclude outsiders from resource expropriation – traditional rules need to be backed by law and state enforcement support.
- Awareness raising: a more generalised transformation of the policy and attitude in support of CBNRM is needed. This requires more awareness among local politicians, and also among many government officials from departments outside of wildlife management who are concerned with resources such as water, trees, soils and grazing.
- Community capacity building: including leadership training and efforts to empower women.
- Giving value to environmental resources: consideration should be given to levies on environmental resources transported out of rural areas, such as firewood, charcoal and thatching grass, which will be charged by and paid to local communities. This right could be made dependent on implementing a sustainable management plan.
- Greater recognition of conflicting interests within communities and a willingness to study and research these: typically, these may be between rich and poor households or between women and men, but can also have other dimensions, such as age, lineage or ethnicity.

- A learning approach in which experience gained in early schemes can be built upon: in particular, new programmes being developed under the environmental banner need to be aware of the experience already gained in past agricultural and integrated rural development projects.

There is scope for building more innovative partnerships with traditional and spiritual leaders over environmental matters; however, this will mean real partnership, rather than using these leaders as a cheap means of implementing an externally set agenda.

Those bringing in participatory techniques need to build up their experience and to be aware that participation can bring great benefits but can also be misused like many other processes.

Part IV

Creating an Enabling External Environment

Chapter 9

Secure Access to Land

INTRODUCTION

In order for farmers to feel that it is worth their while investing time and money in maintaining and building up the resources of their land (such as soil fertility, erosion control structures and trees) they need to feel confident that they, or their families, will benefit in the medium and long term from this effort. This does not necessarily mean that they have to own the land in a legal sense, but they need tenure security over the land and the benefits flowing from their good husbandry. Within the region, land tenure is generally dualistic, with a:

- formal freehold/leasehold system applying to the large-scale commercial sector;
- traditional use-rights system applying to smallholders in the communal sector – within this, arable areas are generally used on an individual or household level, while other resources, such as grazing, tend to be common pool (shared but regulated) or open access (unregulated).

The pressure on land and its associated natural resources differs throughout the region. There is greater availability of land in Zambia, parts of Namibia and parts of Botswana, compared with southern Malawi, many communal areas of Zimbabwe and ex-homeland areas of South Africa. However, in all countries, there is concern that population growth, liberalisation and other changes will undermine smallholders' continued access to land and thus their livelihood security.

LAND SECURITY, ALLOCATION AND ADMINISTRATION

In many countries, there is dual responsibility for land allocation and administration between formal and informal systems (eg by local councils or land boards and local chiefs). In many areas, tenure security of crop land, through traditional systems, provides considerable security, and surveys have shown that smallholder farmers feel they their land rights are secure. However, there is also experience, particularly in areas with patrilineal inheritance patterns, to show that women do not feel they have security of tenure under traditional systems, often losing their land on the divorce or death of their husband, particularly if they do not wish to remarry into their husband's family. Traditional inheritance patterns are not static, however, evidence from Botswana and elsewhere suggests that inheritance may be becoming less discriminatory towards women. A number of local NGOs have been working on gender issues relating to land in both the formal and traditional systems (see Box 3.1).

Experience with mass titling, such as that tried 25 years ago in central Malawi (which was accompanied by a prohibition by the local chiefs on land sales), indicates that the exercise does not necessarily have an impact on security. In some areas, an ongoing land titling process operates alongside traditional use rights – for instance, in Botswana, some smallholders get land allocated (or confirmed) by the land board; others rely on informal rights. This is in part a slow replacement of traditional authority by more formal and perhaps democratic structures. However, typically the larger land holders, and those wishing to invest in water or fencing development, use the land board, creating a dual system which may in future result in a lack of security for those with 'only' traditional tenure. A similar concern is raised in Zambia where the 1995 Land Act enables leases to be granted in communal areas. The costs of securing leases will rule out poorer people, and smallholders with only traditional rights from chiefs might find themselves outflanked. In Zimbabwe, allocation of land rights in communal areas has been removed from *kraalheads* and chiefs and given to elected local structures – VIDCOs. The Land Tenure Commission report (1994) found that the new structures were not particularly democratic, had low local legitimacy and were less accessible to women than the traditional structures. Where a dual system is in place, it is important that there is compatibility, communication and mutual recognition between the two systems. Where a traditional system is being replaced by a titling system, it is important that poorer farmers, and especially women, do not lose out in the transition.

In heavily populated areas, evolution of land tenure is likely to be particularly rapid and important – yet we tend to know very little about what is happening in practice. Issues such as the speed of informal privatisation of land; the emergence of rent; and changes in inheritance patterns (eg from matriliny to patriliny) can be crucial to people's sustainable management and investment in land; such issues need to be better understood.

It is also important to know more, in different localities, about the interrelation between investment in sustainable land productivity – tree planting, contour ridging, stabilisation of gullies, vetiver grass planting – and land tenure security. In some places (eg parts of central Mozambique), investment, such as tree planting, is apparently not allowed by local custom since it implies a longer term claim on land than the permission to cultivate allocated by the chief. In other areas activities such as planting trees and fencing are important ways of staking claim to land (sometimes illegitimately); in still others there are examples of farmers being reluctant to invest in land improvements because of lack of security – with them or their families not being able to reap the benefits.

In order to encourage sustainable agriculture, local custom and practice needs to develop and be supported where necessary by legislation and policy change, so that:

- Investment in sustainable intensification is allowed and encouraged.
- There is sufficient tenure security for households to feel confident that if they invest in their land they will be able reap the benefits of the investment (or be compensated for their investment).
- There is appropriate equity between more and less powerful households, women and men and across the ages to ensure that sustainable intensification is broad based.

SECURITY TO COMMON LAND RESOURCES

While smallholder rights to crop land are still relatively secure, the same cannot be said for grazing and other common property rights. These are generally under greater threat throughout the region as a result of:

- increasing use (commercial exploitation such as charcoal for cities, but also due to increased population), creating unsustainable pressure;
- breakdown of traditional control systems (common pool becoming open access);
- breakdown of government enforcement capacity (concerning tree cutting, cultivating sloping land, etc) without replacement by community-based alternatives;
- privatisation by powerful or rich individuals, both legally and illegally.

For example, communal grazing land is being privatised through fencing or control of boreholes with the legal consent of the land board in Botswana, while in Namibia some people are exploiting the policy vacuum over land tenure to do the same without legal authority. In Namibia land legislation is currently under debate, and although crop land is likely to come under individual or household control, it is not clear whether grazing land will come under individual or exclusive group ownership.

Some form of exclusive group rights is needed for community management of common property (see Table 8.2). The legal basis for community management of common property resources is sometimes unclear – often because of the difficulty of defining a relevant 'community'. In Namibia, there is a legal structure in the form of conservancy for managing wildlife (see Box 8.2), but it does not cover other land resources such as grazing. In Zimbabwe, CAMPFIRE projects exploit flexibility within district council regulations rather than the legal rights of communities. In South Africa, a legal instrument – the Communal Property Association – has been developed to provide a legal framework for community ownership of resources.

Community-based management of natural resources was discussed further in Chapter 8.

LAND SALES AND ENTITLEMENT LEASING

Land sales in communal areas are not allowed in most countries in the region, although there is growing evidence that they are, in fact, occurring, leading gradually to the development of a land market and ultimately to the privatisation of communal land. The effect of an official or unofficial land market can have both positive and negative impacts on smallholders.

Table 9.1: *Interaction between Land Market and Smallholder Opportunities*

Land Market	Advantages	Disadvantages
None: (land sales not allowed)	• Poorest farmers likely to retain access to some land, even if insufficient for survival. • Land may be available for those without access to family land, especially young people and divorced women.	• Poorest farmers unable to realise capital from unviable holding, or when they wish to migrate to town or to another area. • Credit may be restricted since land cannot be used as collateral.
Yes: official or unofficial	• Capital can be realised to facilitate move to more productive area or move to other employment. • Land could in theory be used as loan collateral.	• Households can be made landless through non-payment of debts or forced sale in crisis situations. • Young people and divorced women may not be able to get access to land.

Traditional tenure systems are not static but are evolving, particularly as pressure for land increases. The direction of this evolution, and in particular how it affects issues of sustainability and access to land by women, young people and the poorest households, needs monitoring and, if necessary, influencing. Those working to protect smallholders' access to land may need to make a strategic choice between fighting to maintain communal tenure, and accepting that privatisation is inevitable and working to ensure that poorer smallholders get access to privatised land.

In some communal areas with incipient land shortage, well-connected households have had more land allocated to them than they can currently cultivate, which they seem to be holding for their children. While, on the one hand, this is a good example of an investment in sustainability, it also means that other households, who are currently desperately short of land, are having to cultivate without fallows alongside this currently unused land. This phenomenon is probably new in that many traditional tenure systems have not had to operate in a situation of land shortage before. Custom and practice need to evolve to deal with these new realities, with the state and NGOs supporting communities to develop equitable and sustainable procedures.

There are various possible compromises between the current system of prohibiting land sales and a free market approach. One possibility is entitlement leasing, in which a household's entitlement to land (or another resource such as grazing) is rented to others. Practical experience in Southern Africa of developing this approach to support sustainability is very limited. There is evidence that crop land in communal areas throughout Southern Africa is being rented on a limited and informal basis (sometimes the payment is symbolic) and that such leasing seems to be increasing. No examples were encountered of other entitlements being leased on an individual basis, although some CBNRM projects are examples of communities leasing their entitlements to a third party. Some carefully designed action–research projects are needed, ideally as part of broad-based CBNRM programmes.

Land Tax

Land taxes are being advocated in the region for a number of reasons:

- to discourage hoarding and inefficient use of land by large farmers;
- to bring more (previously underutilised) private land onto the market, thereby lowering land prices and facilitating land redistribution;
- to provide more funds for local government.

Land taxes (or lease payments to the state) exist in several of the countries for leasehold areas (which tend to be occupied by larger commercial farms). In general, lease payments are low and sometimes are not paid.

For instance, in Botswana 6400 hectares of tribal grazing-land policy ranches are rented out at very low rates and in Malawi it is recognised that many estates do not pay their leases. Although land taxes can contribute to improved equity and, in the right circumstances, sustainability, there are some issues that need to be considered:

- Land taxes need to be structured to reflect the production potential of different land and should not be an incentive to bring land into unsustainably intensive use.
- If communal areas are excluded from land taxes, this might discourage richer farmers from moving from overcrowded communal areas to leasehold or freehold areas – this move of richer farmers, often with large numbers of cattle, from communal to private land areas has been a particular policy objective in Namibia and Botswana in order to relieve pressure on communal areas.

LAND REDISTRIBUTION

In areas with particularly inequitable land distribution and acute land shortage, such as southern Malawi, Zimbabwe and South Africa, land reform is seen by many as a political and economic necessity. As a highly politicised issue, programmes are likely to be driven by political expediency rather than by sustainable principles; this makes influencing programmes in favour of sustainability and equity particularly important.

Experience of large-scale land redistribution in the region is still fairly limited. The largest and longest running programme has been in Zimbabwe, where 71,000 families have been resettled and a further 20,000 have benefited from access to increased grazing. This is a significant number, but still represents under 10 per cent of communal area farmers over 17 years, meaning that the process has not even kept pace with population growth. Although popular opinion in Zimbabwe tends to be critical of the resettlement programme, reports by the Comptroller and Auditor General (1993) and by ODA (1988) were generally positive about the impact on those resettled and the value for money (ODA, 1996). The rate of resettlement has slowed down since the mid 1980s, leaving overcrowding in many communal areas to rise rather than fall; cynicism has tended to be fuelled by the fact that some of the appropriated land has been used by rich farmers for considerable periods rather than being redistributed. A new government policy paper in 1996 made it clear that, even if the resettlement programme is completed in full, a substantial large-scale commercial farming sector will remain, and the number of households remaining in communal areas will be several times higher than those that will have been resettled.

The experience in Malawi over the last 20 years has been in the opposite direction – with 20,000 new private land holdings being carved out of communal areas; much of this acquisition has been speculative, with the land currently used at low intensity, while the diminished communal land is used ever more intensively. The net effect has been detrimental to both equity and sustainability. Land reform in South Africa is in its very early stages. In one respect, the South African land reform process is more bottom up than in Zimbabwe since individuals can get grants of up to R15,000 (US$2500) to purchase land. Therefore, the process is more demand driven – with a number of households banding together to purchase a farm, they are, in theory at least, more involved in the process of selecting suitable land and deciding how to divide and manage it (see Box 9.1).

There are a number of cross-cutting themes emerging from the different land reform programmes in the region, as described below.

Equity versus Productivity

The argument is whether preference should be given to those in greatest need or to relatively better-off (larger-scale) smallholders who are considered better able (by some) to use the land productively and with less support. There is, however, very little evidence that better-off smallholders actually do use their new land more sustainably and productively than the very poor – more important is likely to be the way the resettlement programme is designed and managed. Evidence from Zimbabwe suggests that smallholders, in general, use their new land very productively in comparison to medium- and large-scale commercial farmers (Land Tenure Commission, 1994). Despite this, the new resettlement policy (Government of Zimbabwe, 1996) indicated that in future land redistribution will meet the needs not only of landless people from overpopulated wards, but also:

- successful peasant farmers who want to venture into small-scale commercial agriculture; and
- indigenous citizens who have the means and resources to enter into large-scale commercial agriculture.

There is, therefore, a danger that a process, already unable to meet the needs of the poor, will be diluted to allocate land to the better off.

Stakeholders in Sustainability

Use of redistributed land has tended to be a rather top-down process, with the state trying to police sustainability and productivity. Problems are often attributed to the 'state not doing enough'. However, for programmes of the scale needed, the state will never have the resources to 'do enough' – therefore, to be practical, a bottom-up process is needed, with greater responsibility given to farmers. This makes it essential that the framework within which the land is managed provides the right incentives for sustainable use, and that farmers are given training in communal management (see Table 9.2).

It is important for participants in land reform to have a vested interest in the land's long term sustainability and to have the capacity to manage the process. This requires:

* reasonable tenure security (some programmes are caught in a Catch 22 situation – wanting to have the sanction to evict settlers if they do not farm 'properly', which means there is little security, and therefore little incentive for long term investment);
* appropriate technical support;
* appropriate institutional arrangements, particularly over the management of group property resources such as grazing (including avoidance of unmanageably large groups).

Cousins (1995) proposes a checklist of factors that should be taken into account in the design of common property regimes in land redistribution programmes; see Table 9.2.

Sustainable Plot Sizes

In South Africa, doubts are being raised as to whether the R15,000 (US$2500) grant per household is sufficient to purchase enough land to sustain a household. There is, however, a danger of being oversimplistic about viable holdings – rural households throughout the region derive part of their income from off-farm sources (this can be a source of investment in the farm enterprise, as well as risk-reduction through diversity). There is no reason why land should be used less sustainably because it supplies only part of the household livelihood (see Chapter 3). A greater problem is that, because of the scale of the grant, the size of farms for sale and anti-subdivision legislation, often rather large groups end up buying a farm together. This makes sustainable resource-management difficult. Another issue in many resettlement schemes is the provision of plots for the second generation.

Individual settlement schemes each have their peculiar histories, have distinctive pieces of ground and involve diverse communities. The case for participatory planning is very strong, with national programmes needing to move away from a blueprint approach. In Zimbabwe, there are some more recent pilot approaches to land reform, involving greater participation; these may become a model for the future. The case study from South Africa illustrates the complexity, with different groups of settlers having different priorities, which only bottom-up, case-by-case planning and negotiation can resolve (see Box 9.1).

Table 9.2: *Checklist of Factors for Design of Common Property Regimes in Land Redistribution Programmes*

User group issues	• Have rules for user group membership (entry and exit) been clearly defined?
	• Is the size of the user group appropriate, in relation to the resource base?
	• Do institutional arrangements and/or organisational structures provide a voice for the less powerful within the group?
	• Do institutional arrangements promote the emergence of a community identity?
Resource management rules	• Do rules clearly establish the conditions for collective decision-making over resources (eg the right of the group to establish limits on individual use)?
	• Have juridical boundaries been clearly defined? In non-equilibrium ecosystems, have boundary issues been sufficiently clarified?
	• Are operational rules easy to understand, unambiguous and easily enforceable?
	• Has the number of rules been kept to a minimum?
	• Do rules make provision for monitoring and punishment of infringements?
	• Do rules take into account potential conflicts between different users of resources, and between different categories or groups of users?
	• Do rules establish the organisational form for decision-making (eg elected committees)?
	• Does the user group have the right to modify and adapt the operational rules?
Authority and enforcement	• Has authority been allocated at the appropriate level(s)?
	• Have relationships between user group and government agencies, legal and customary authorities been clearly defined?
	• Do institutional requirements have a recognised legal identity?
	• Have mechanisms been designed for negotiation, mediation and conflict resolution, within and between user groups?
Resources	• Do partitioning rules take adequate account of ecological and technical realities (eg key resources; the feasibility of fencing)?
	• Is there sufficient flexibility over boundaries in non-equilibrium systems?
	• Do rules take into account the spatial and temporal variability of resources?

Source: Cousins, 1995

BOX 9.1 LAND REFORM CASE STUDY, PAKKIES COMMUNITY, SOUTH AFRICA

This case study highlights different priorities amongst members of the same community in selecting land for settlement under the Land Reform Programme.

The Pakkies community lives on two adjoining black-owned farms in the Kokstad district of KwaZulu–Natal (these farms are a historical relic; the current owners are the descendants of the original owner, while most other black owners lost their land to white farmers decades previously). Many of the residents are tenants of the current owners and are farm workers who have been displaced from large-scale, white-owned farms in the district. The land is used for maize and to graze cattle and goats. The land owners have depended more upon the rentals paid by tenants for their income than on agricultural production. Livestock owned by tenants has increased pressure on the grazing resource, leading to its deterioration. The Pakkies community has received no services from the Department of Agriculture (DoA) in recent years, and no controls under the Soil Conservation Act have been applied.

In the period leading up to the 1994 elections, a rent boycott on the part of tenants led to severe friction with the landowners, who brought an eviction order on them. In 1996, the Department of Land Affairs was requested to make land available to those without land, utilising the R15,000 (US$2500) household grant. A delegation from Pakkies visited a number of farms in the area to assess which might be suitable for purchase.

In March 1997, a participatory workshop was held by the Department of Land Affairs to assist the community in making an informed decision about the land to be purchased and how it would be used (Cousins, Joaquim and Oettle, 1997). The department was concerned that land which had been bought elsewhere with the R15,000 grant did not provide the basic necessities for sustainable livelihoods, and that this was resulting in degradation of the resource base (Mdu Shabane, pers comm). However, it was felt that if the potential beneficiaries could be assisted to think through the options available, and consider how the project would improve the quality of their lives, the chances of establishing sustainable livelihoods would be increased.

By the end of the workshop it was clear that there was a strong group (approximately 90 households) who preferred acquiring land adjoining the existing settlement. They would be able to stay in the neighbourhood and use the schooling facilities available. Many of this group were employed in the area, or in nearby Kokstad, and their primary objective appeared to be a secure and familiar place to live, with access to services. The farm Abergeldie, adjoining the Pakkies land, was the more reasonably priced of the two available. This farm would offer 0.2 hectares of irrigated land per household, a further 0.2 hectares of arable land, and 7.3 hectares of grazing land per household (sufficient for 3.3 LSUs). Limited opportunities for agricultural intensification did not seem a disincentive to this group, who appeared to be more set on acquiring secure tenure than in maximising agricultural income.

Another well-motivated group of approximately 120 households wanted to purchase a farm which would offer real opportunities for agricultural livelihoods, and identified the farm Aloekop as their priority. The farm was well developed, and the grazing land well camped and conserved and highly productive. The group was convinced that the farm offered them good potential for agricultural livelihoods, including potential for irrigated high-value crops such as vegetables. An initial survey shows that the Aloekop farm would provide 120 households with 0.55 hectares of irrigated land, a total of 1.77 hectares of arable land and 8.4 hectares of grazing (sufficient for approximately four LSUs) per household. Since all of the arable land is potentially irrigable from water sources on the property, the potential for intensification under irrigation is excellent, given relative ease of access to markets for vegetables and fruit.

Additional capital and appropriate support services will have to be available to ensure that the beneficiary community is able to maximise its opportunities for agricultural livelihoods in the long run. Furthermore, given the relatively large size of the beneficiary group (not all of whom know each other), and the fact that most will be moving into a new environment, sound institutional arrangements will be critical. Devolution of responsibility for resource management and conservation to smaller user groups (for example, groups of graziers allocated to specific areas of the range land) can address the problem of group size.

These two groups have a common desire to gain secure tenure, but the latter prioritises land-based livelihoods, while the former sees land-based livelihoods as important supplementary sources of income, though second best to off-farm employment opportunities.

Lessons learned

- Different potential land recipients have different needs and aspirations. Therefore bottom-up planning is needed from the start to enable appropriate land selection.

- Even within a limited geographical area and with the limited budgets available, there is scope for strategic choices to be made about which resources to buy.

- Shortage of community organisation skills rather than technical farming skills is likely to be the key limiting factor, both among the smallholder farmers and those available to provide support (eg within the DoA). Initial project design, with subdivision of key resources into smaller units, managed by smaller, more cohesive groups of households, may be crucial factors in determining sustainability.

- Experienced community facilitation and participatory planning skills are needed; these may be best provided by a specialist NGO.

Source: EDC, 1998b

Conclusion and Recommendations

Land tenure security is an essential incentive for farmers to invest in long term sustainability. However, this does not necessarily mean individual freehold ownership; there are a variety of different secure tenure possibilities. Traditional forms of tenure are not static but are evolving in response to new realities. The direction of this evolution can be influenced, and communities need to be supported to develop systems which encourage sustainability and access to land by women. Similar advocacy for sustainability, women and the resource-poor is needed if systems are to change from traditional to formal (eg titling) arrangements. Common property rights are under threat in many areas and community authority needs to be reinforced, with recognition and legal back-up for exclusive group rights. Often, too little is known about current land tenure, how it evolves and its interaction with sustainable management.

A market for crop land is developing in some areas, bringing both advantages and disadvantages. Perhaps this is inevitable and, if so, needs to be managed in a way that the threats to women, the poor and sustainability are minimised. Various compromises may be possible, including possible leasing arrangements.

The lessons from the land reform programmes in the region are:

- Settlers need tenure security.
- A more participatory planning process is needed, taking into account diversity of objectives and circumstances.
- More attention needs to be paid to institutional development, especially for the management of common property resources; overlarge groups should be avoided.

Chapter 10

Agricultural Extension and Training for Sustainability

INTRODUCTION

There is a widespread view throughout Southern Africa that agricultural extension has underperformed, especially in relation to resource-poor smallholders. There is less agreement on the reasons for under-performance, and some of these are clearly country specific. Common biases against sustainable smallholder agriculture in research were given in Chapter 4.

In some Southern African countries there is a crisis in agricultural extension due to lack of money; however, lack of funds is not the whole story, as Table 10.1, which is a brief analysis of some of the reasons given for under-performance in one relatively well-resourced extension service, shows.

The changes needed for research and extension in the region are divided into:

- approach/philosophy (what needs to be done);
- new strategic alliances (who does what);
- divestment (what not to do);
- management (doing less better);
- training (having the capacity to do it).

'WHAT NEEDS TO BE DONE' – APPROACH/PHILOSOPHY

Extension services in Southern Africa are still built on the foundations of a transfer of technology approach (see Chapter 2). All are, however, under-going change with:

Table 10.1: *Common Reasons Given for the Low Impact of Agricultural Extension in Botswana*

Reason for Failure	Comment
Not enough front-line extension workers.	Not really a viable reason, given that Botswana has one extension worker to 300 families compared to an average for sub-Saharan Africa of one to 1800 families. The lack of suitable transport for some extension workers has, however, reduced their effectiveness.
Insufficient or inappropriate training for technical assistants.	Training facilities at the Botswana College of Agriculture (BCA) have improved markedly in recent years; qualifications on entry (Cambridge) and length of course have also increased and compare favourably with other countries in Africa. However, the course has a 'modernisation/technical' focus, with little emphasis on farmer participatory techniques, indigenous knowledge and the role of the extension worker as facilitator. Both high external-input and low external-input agriculture is taught at BCA, but there is more emphasis on high-input farming.
Inadequate selection and supervision of extension workers (EWs).	Many farmers complain that the EWs exist but they never see them. Both selection and supervision of EWs is very centralised, with virtually no farmer or farmers' committee participation.
Inadequate improved agricultural techniques to promote.	The Division of Agricultural Research and the ALDEP (Arable Lands Development Programme) have consistently claimed that they have developed varieties and techniques that will significantly improve yields. However, many of these have not been enthusiastically taken up by farmers. It is probable that many of these techniques are not actually as advantageous under family sector conditions over a range of rainfalls as the researchers believe.
Drought.	From one point of view, it is legitimate to blame drought for the failure of crop agriculture and of programmes to improve it. On the other hand, given the drought-prone nature of Botswana, the challenge is to develop agricultural systems and services that work in this drought-prone environment.
Extension workers spending time on drought relief rather than on agricultural extension.	This is true, as many EWs have spent 80 per cent of their time in recent years supervising drought-relief programmes (measuring fields for destumping and ploughing grants).
Farmers not serious about crop agriculture.	Although this may be true for some farmers, and many of the youth, other farmers are extremely dedicated and rely heavily on crop production.
Wrong overall approach by the Ministry of Agriculture.	Farmers are seen as the problem rather than as the solution. A top-down, technical, teacher-pupil system is promoted, with little more than peripheral farmer participation.

- increased allocation of research resources towards smallholder priorities, more on-farm research, recognition of indigenous technical knowledge (ITK), etc;
- modifications to training and visit (T&V) extension systems, towards more flexible approaches with more farming system emphasis.

Yet, fundamental change has generally not taken place – elements of farmer participation have been added on to what is still essentially a top-down approach. Lack of fundamental change achieved so far should not necessarily be taken as a criticism; change will take time. What is important is that the process of change occurs in the right direction, even if progress often appears slow. Fundamental change is not easy; even if some sectors within the agricultural establishment want to change, they may be blocked by others who feel threatened by a more farmer-driven approach (EDC, 1997d; Chambers, 1997). An additional difficulty in managing the change in approach is that it is compounded by changes forced on research and extension, as a result of budget cuts and economic liberalisation.

Key roles for extension at the community level should be:

- Support of a learning environment among farmers by providing technology choices and information, encouraging and supporting farmer experimentation; and facilitating farmer-to-farmer exchanges and visits.
- Facilitating group/community capacity, where agricultural development requires group or community actions.
- Networking – facilitating contact between farmers and their organisations and other entities (commercial companies, NGOs, funders, research and policy-makers).

Some points to note:

- This facilitatory role does not preclude the transfer of technology (however, the transferred technology is chosen by the farmer within a learning environment).
- This role does not necessarily have to be played by one person or organisation (it can be played by a multidisciplinary team from a variety of organisations).
- This role does not necessarily have to be implemented by government staff (however, the public good element of much research and extension indicates a continuing role for public-sector funding).

Decentralised, facilitatory extension needs more skills, different skills and better management than top-down, centralised, technology transfer. There needs to be more flexible, better-trained and more community- and gender-aware teams and networks who are able to respond to changing circumstances. This has implications for collaborative alliances and management.

Some of the recent debate on extension in the region concerns inter-mediation – whether to work directly with individuals, or through master farmers, contact farmers and existing groups – or whether to form specific extension groups. Extension designed to support sustainable smallholder agriculture will need to respond more to local circumstance, which means being more flexible about whom to work with and how. There will need to be affirmative actions to make sure the poorest and women farmers are reached; without this, experience has shown that more attention is paid to the better-off farmers, and approaches develop which are less appropriate to poorer and women farmers, creating a vicious circle. Extension provided by a diversity of organisations is probably more likely to be able to provide the flexibility of provision required.

Structural change alone may not be successful, unless it is accompanied by changes in attitude and approach. This was experienced in Malawi, when a very successful NGO programme, performing extension with church women's groups, was adopted as a component of national extension strategy. The government extension workers found it difficult to adapt to working with groups, which they did not control (see Box 10.1).

'WHO DOES WHAT' – STRATEGIC ALLIANCES

The new plural environment requires a re-thinking of who does what and what sort of collaboration between different players is needed. Increased commercial involvement in supply and marketing provides opportunities for a range of extension and commercial interactions such as:

* extension workers facilitating contacts between sales representatives and farmers (eg invitations to open days, facilitating contact with farmers groups, etc);
* extension workers promoting products that are sold commercially, implicitly, or perhaps explicitly with sponsorship deals (at the most superficial level, many extension workers already wear hats and t-shirts with brand names of various inputs);
* sales representatives providing extension information or demonstrations alongside their product;
* extension workers facilitating marketing contacts for farmers.

There are both opportunities and dangers in this collaboration. Clearly, cost sharing with the private sector can increase sustainability, but dangers include bias towards purchased inputs, bias towards specific products, and corruption. In Zimbabwe, input suppliers are working alongside extension at a local level to mount demonstration plots (see Box 11.3); while this is fine in itself, the demonstrations cover the use of inputs, but not long term, low external-input alternatives – thus reinforcing existing biases. Guidelines need to be developed for extension collaboration with commercial companies. These are probably best developed by a working

---

Box 10.1 Agricultural Extension with Church Groups in Malawi

In the early 1990s the Malawi government agricultural extension staff worked almost entirely through credit groups, who were borrowing money for fertilizer in order to grow maize. These groups tended to be male-dominated and involved the slightly better-off farmers. This meant that the vast majority of poorer and women farmers were by-passed by the government system.

In 1990, a church-linked NGO, the Christian Service Committee (CSC), started a pilot extension programme in which simple, low-cost and sustainable techniques were promoted by specially recruited extension workers, working specifically with church and other existing community-level groups. It became apparent that these CSC extension workers (most of whom had previously worked for the government extension service) were able to:

- reach a larger number of farmers;
- reach a higher proportion of poorer and women farmers;
- successfully introduce appropriate techniques to these farmers;
- persuade some keen farmers to become 'extension multipliers' to spread the messages still further.

In reaction to this programme, the government extension programme officially adopted a policy of working with existing community, church and mosque groups in 1994. The results, so far, have been somewhat mixed, with many government extension workers finding it difficult to relate effectively with these existing groups. The problem seems to be that the government extension workers were used to working with 'their' credit groups, where farmers were obliged to come to meetings and adopt recommendations in order to get credit. Many extension workers feel uneasy about working with groups in which they are 'guests' and where they cannot compel farmers to follow their advice.

A lesson learned has been that both the policy governing working with differ-

---

party, including representatives of government, commercial companies, farmers and NGOs.

There is also plenty of scope for more collaboration between the different arms of government impacting on different aspects of sustainable rural livelihoods, such as agriculture, forestry, the environment, land, water, commerce and education. Ideally, this collaboration needs to be encouraged at all levels but is particularly important at the local level, where effective collaboration between different field workers can make major improvements to overall performance. The potential for collaboration is large; for example, in just the case of schools and extension, this could involve:

- local collaboration between primary and secondary teachers and extension staff, to make agricultural teaching more closely related to

the agricultural developments going on in the local community, and better compatibility between information given to children and to their parents;
- use of schools as a resource for local farmers (eg use at weekends and holidays, visits to school vegetable gardens, schools as sites of tree nurseries, etc);
- some continuity of contact with school leavers, with a 'hand-over' from the teacher to the extension worker when pupils leave school.

NGOs have often led the way in the region in developing more participatory methods, in critiquing conventional approaches to landuse and in piloting various approaches to sustainable agriculture. There is considerable scope for more strategic collaboration between government and NGOs, and it is disappointing that so much collaboration is still ad hoc. One probable reason for this is that there is still a mind set that the natural state of affairs is:

- permanent government extension;
- temporary NGO agricultural projects (often running until government has the capacity to take them over).

This needs to be challenged by a future vision of locally based NGOs and CBOs having a permanent (but evolving) role in community-based extension, with funding coming both from local government and the local communities.

The relationship between NGOs and ministries of agriculture can take many forms and has been reviewed by Wellard & Copestake (1993). A typology of different relationships is given in Table 10.2.

Some of the conflict between government extension services and NGOs may diminish as government services adapt to a more plural environment, and when all those involved recognise that extension is more about providing choices to farmers than specific recommendations.

In many countries in the region there has been a shift in resources from local government to NGOs; this is particularly the case where international donors have lost confidence in funding government directly and therefore channel increased funding through NGOs. This has in some countries created a very visible resource imbalance between NGOs and local government (eg with NGOs having an excess of four-wheel drive vehicles and local government having none). In some cases, positive working relations have been maintained between NGOs and local government, with the NGOs contributing to local government capacity building; in other cases, the result has been jealousy and conflict. In future, it would be more healthy if the situation was at least partly reversed, with more equal resource bases and a leaner but better-resourced local government providing some funds to local NGOs and community organisations, perhaps on a contractual basis for implementing certain defined services.

Table 10.2: *Typology of Relationships at Local Level between Government Extension and NGOs*

Relationship	Main Features
Conflict/ suspicion	The government may be threatened by: • any alternative extension to farmers; • alternative recommendations to the government orthodoxy (NGOs are acceptable if they toe the line). Similarly, NGOs may oppose government services because they consider them to be giving incorrect recommendations (typically, in favour of high-input agriculture).
Ad hoc collaborative	Typically, the extension service provides ad hoc technical input into NGO-organised programmes and courses, and the NGO provides occasional transport and training for government extension workers.
Minimalist coordination	Government may host information-sharing meetings for the different NGOs working in the district. There is some sharing of fieldwork plans and attendance at planning meetings, but there is no joint strategic planning to exploit the possible synergy between the two organisations.
Counterpart	NGOs often have explicit or implicit plans for the government to take over the programme when they withdraw (eg Namwala Cattle Clubs – see Box 7.1). Expectations are sometimes unrealistic or badly planned.
Catalytic	The NGO has an agenda (sometimes hidden) to improve the government extension through methodology development, training and resource sharing.
Contractual	Government may provide funding for NGOs to do specific extension tasks; or within a donor-funded project different activities may be contracted to NGO and government for implementation (see Box 10.2 and PDT below).
Strategic collaboration	Government extension and NGO(s) plan and work together on the basis of their different comparative advantages; this remains rare.

This is likely to require increased donor assistance to local councils in the short and medium term.

NGOs are rightly criticised by many in government for their lack of accountability to the local community and their opportunistic approach, setting up projects where convenient without necessarily working in a cost effective or replicable way. To play a longer term role in agricultural development, many NGOs will need to:

• Develop local structures which are accountable to local communities.
• Develop systematic and objective systems of monitoring their performance.

Donors, many of whom have switched funding from government to NGOs in recent years, need to be more rigorous in insisting on accountability and effective performance.

There seem to be few examples of government contracting out of either research or extension to the private sector or to NGOs. In Mozambique, the secondment of government extension staff to NGOs has been quite common; the government continues to pay the person but the NGO provides management, transport and often a salary top-up. An example of contracting out or collaboration at the local level is in the Tswapong Hills area of Botswana, where the government's Arable Lands Development Programme (ALDEP – see Box 11.12) is largely implemented by a local NGO, the Palapye Development Trust (PDT). There is some preliminary discussion of contracting out extension to farmers' organisations in Namibia (see Box 10.2).

With the expansion of contract-growing schemes with smallholders, especially for cotton (see Box 11.10), there is increasing experience of extension providing advice to farmers in such schemes. Often the relationship between the outgrower scheme and extension is fairly informal, with extension workers giving advice to scheme members as part of normal services. There is, however, interest in some quarters to extract cost recovery from this sort of involvement.

BOX 10.2 FARMER-DIRECTED EXTENSION SERVICES IN NAMIBIA

One area of reform which is being debated amongst communal farmer associations in Namibia, but which has yet to impinge on the thinking of the government's extension planners, is the question of how to make extension services more answerable to farmers. Extension services are run in a highly centralised way, with little influence exerted on them at the regional, let alone the community level. Farmer associations (and other local bodies such as regional councils) are vociferous in their views that, partly because of this, extension services are failing to meet farmers' real needs. A number of farmers' associations claim that, because they represent and are accountable directly to farmers, they will be in a good position to manage local extension services. These proposals, although radical, fit in with the government's overall policy of contracting out services to the private sector.

Source: EDC, 1997d

'WHAT NOT TO DO' – DIVESTMENT OF NON-EXTENSION DUTIES

Throughout the region, extension has been burdened with a variety of duties which are now considered better performed by others or which have detracted from their core activities. In Botswana, drought relief has been a major task, taking up to 80 per cent of extension's time in some years. In Malawi, before the collapse of the credit programme, extension workers spent nearly all their time working with the 20 per cent of the (generally) richer and male smallholders, who were members of credit clubs (see Box 11.6). The worker could ensure attendance at meetings and adoption of recommendations by controlling who received credit, and some have found it difficult to adapt to working in less coercive ways (see Box 10.1).

Economic liberalisation has led to government withdrawal from many activities now considered to be better handled by the private sector. In Namibia, extension was involved in providing a range of services, including ploughing and input supplies, which it has struggled to divest (see Box 10.3). An interesting condition of the divestment was to start an input voucher scheme, thus giving a new input security 'safety net' role to agricultural service provision. Those concerned, however, would like to see extension's role in administering the voucher scheme as minimal, and the main role going to some other agency, probably local government (see Box 11.12).

BOX 10.3 DIVESTMENT OF NON-EXTENSION DUTIES IN NAMIBIA

In Namibia, since independence, the extension service has provided a number of services, such as highly subsidised tractor hire, and the selling of fertilizer, seeds and tools – which has served the interests of a small, but vociferous group. Managing these services has been an overwhelming burden for the three critical months at the start of each cropping season and has limited other activities of the extension services. The decision to divest the extension service of responsibility for operating these services is considered a critical turning point for developing extension services. It was brought about by the converging interests of agribusiness – which saw government services as unfair competition, preventing them from providing commercial services – and extension managers, who saw that continuing responsibility for tractor hire services would constrain the development of other extension services. The process involved contracting independent research to analyse the implications of the various options for commercial agricultural services. These researchers were guided by and reported to a large task force, comprising a wide range of interested parties, most notably those representing communal farming and agribusiness interests.

In 1995, the cabinet approved the task force's proposal, under which the Directorate of Extension and Engineering Services would withdraw from the direct provision of ploughing services and the retailing of fertilizer and seeds, as and when the capacity of the private sector had developed to the stage when it could provide these services efficiently. The scheme involved training programmes for, and the sale of tractors to, the private sector. It also proposed the introduction of a voucher scheme to subsidise the purchase of ploughing services, seed and fertilizer for the poorest farmers through commercial channels (see Box 11.11).

Source: EDC, 1997d

'DOING LESS BETTER' – IMPROVING MANAGEMENT

Management of government extension has been identified as a constraint in all countries in the region, and improvement in management is seen as a major opportunity for improving extension performance. Some donors are shifting emphasis from providing technical support to providing organisational and management support to ministries of agriculture (eg GTZ in Zimbabwe).

In most countries, management problems are part of wider problems within the civil service:

- low pay, low morale, few resources to do the work;
- inflexible structures – difficult to hire and fire and reward on merit;
- loss of better-educated, experienced or motivated staff to private or NGO sectors.

While further discussion of these issues is beyond the scope of this book, it is emphasised that these difficulties have profound implications for agricultural services and policy delivery.

Agricultural extension is particularly difficult to manage – impact is difficult to measure and the extension worker's job is isolated and erratic; therefore, opportunities for avoiding work are common (unlike, for instance, a teacher who is supposed to be in class X teaching syllabus Y at time Z). In the past, the difficulty of managing extension was tackled by making the organisation more rigid (supply led), the classic example being the T&V extension approach. This conflicts with the more flexible, demand-driven approach increasingly being recognised as appropriate.

There are numerous projects in the region, many with strong donor involvement, trying to improve research and extension. Typically, these projects have one or more of the following objectives:

- changing the approach to extension or research by demonstrating new ways of working, often through pilot projects;
- strengthening the capacity of research and extension in specific areas or approaches;
- getting something done by avoiding existing bottlenecks in research and extension through a project with more flexible rules, a separate budget system, and incentives for staff.

Some of these projects are an evident success – for instance, collaboration between the Ministry of Agriculture and a foreign university, in the Malawi Agroforestry Extension Project, is having a major impact on agroforestry development in Malawi, and in its second phase is developing more into a facilitator and a resource for the whole range of other organisations working on agroforestry (EDC, 1997b). However, there are also many projects of this type that achieve something while running but have little sustainable impact. There are, therefore, some general concerns arising from this approach:

- The existence of a large number of projects within the Ministry of Agriculture means that they can become unmanageable and priorities become donor driven (sometimes called the 'projectisation' of the Ministry of Agriculture).
- Although pilot projects may produce successful results initially, they are often not replicable on a wider scale, or sustainable once the donor/NGO/consultants have withdrawn. This is often because the underlying management problems have been avoided rather than addressed, or the resources available to the pilot (with skilled and motivated staff often being as important as money) are not available on a wider scale.

In order to address some of these concerns there are now a number of restructuring exercises in ministries throughout the region, often coming under the heading of agricultural sector investment programmes. In Zimbabwe, an intermediary step is taking place – the Agricultural Sector Management Project, aiming to create the demand-driven management systems considered necessary for sustainability. This project involves:

- a move away from a large number of disparate projects within the ministry, supported by different donors, to a more coordinated process;
- a move away from top-down, supply-driven service provision to a stakeholder, consultative system with management systems more responsive to stakeholder demand and more results orientated;
- prioritisation of funds for core functions; streamlining; reduction in overlapping responsibilities and exploring the potential for more cost recovery for some services;
- more planning in the context of a plural service environment, in which the commercial sector, NGOs, universities, and international and regional research organisations will all play a role.

The question for those concerned with resource-poor smallholders and sustainability is: *whose demand and for what?* While improved, demand-led management can provide the pre-conditions for supporting sustainability, this will not happen automatically; *there will need to be affirmative action in support of long term sustainability and poorer smallholders to counteract existing biases.*

Decentralisation may give district agricultural officers (or their equivalent) more power and flexibility over staff and budgets to manage effectively. The trade-off is that they need to be made accountable for their unit's impact. This needs to be judged not just on production, but also on sustainability and equity grounds. Therefore, performance indicators need to be developed that measure impact on sustainability, and whether poor and women smallholders are being reached. Farmer, and farmer association, involvement in the selection, management and assessment of both individual extension workers and district teams could play a part in this process (on the basis that they should know if extension is doing a good job or not). Since local situations vary, the solutions will need to vary – what is general, however, is the urgent need to improve management.

'HAVING THE CAPACITY TO DO IT' – AGRICULTURAL TRAINING

Several trends can be observed within formal agricultural training in the region:

- Increasing academic levels – many colleges are increasing their entry level requirements, courses are being extended (often from two to three years), and exit qualifications are being increased (eg in both Botswana and Zimbabwe).
- Whereas previously students had guaranteed (and sometimes obligatory) jobs within the state sector (extension, parastatals, etc), this is no longer the case. Colleges now need to produce students for a diversity of employment – as farmers, as employees in commercial companies, or as NGO workers, etc.
- Although some changes have been made, curricula remain heavily biased towards agricultural yield maximisation using external inputs. Smallholder and farming system approaches, low external-input technologies, participatory techniques, gender issues and communication skills tend to be fringe subjects or are undervalued. Practical work with small farmers as part of the course is quite rare.
- Student-centred training is rare; however, this is an important role model for students who need to promote farmer-centred learning on leaving college.

Scientists love things 'clear' and they have been the ones who have been at the heart of developing and trying to spread technologies – but the process of facilitating technology development, adaptation and spread at community level is 'messy'. It is in fact not a science, it is an art. Many of the most successful community extensionists we know are artists, not scientists. They need to be brought down to earth from time to time with science, but the fundamental reason for their success is approaching community development (of which sustainable agriculture is a part) as an art not a science. They refer to and use science, but their starting point to work and interact with communities is art. For years we have been releasing extensionists into the field who go out as

> *scientists... lots of wonderful information, but it is no good*
> *when faced with the art of community development, the art*
> *of organisational development, the art of culture etc.*
>
> R Librock & J Wilson, pers comm

In-service training can be used as a way to redress past failures in agricultural training (see Box 6.1). This is important and needs to be encouraged, with some provisos:

- Short courses need to be monitored for cost effectiveness; some one-off courses with outside facilitators or short courses that are run in Europe and America can be phenomenally expensive and can only be justified if they have a catalytic or multiplier effect.
- There must be an adequate balance between theory and practice: extension workers need to learn more by doing and by having real contact with resource-poor farmers.
- Training needs to be linked to other institutional changes, especially at management level, within the extension service. There is little point in people returning from training with new enthusiasm to a restrictive environment in which they are not able to practise what they have learned.

CONCLUSION AND RECOMMENDATIONS

Government agricultural extension has tended to suffer from the same biases as research – favouring slightly better-off farmers and simple yield maximisation rather than a range of choices more relevant to resource-poor farmers. The very large gap in all countries between extension recommendation and small farmer practice is an indication of poor research and extension performance. Some of the changes needed are:

- Affirmative action in favour of sustainability through demonstrating and promoting a combination of resource-conserving practices.
- Creation of a learning environment at the smallholder-extension interface – this implies the promotion of choices, the encouragement of farmer experimentation and the facilitation of farmer-to-farmer and farmer-to-research exchanges.
- Decentralised organisation to give local managers more flexibility to organise extension, according to local conditions – the trade-off would be that local managers would need to be accountable for the local impact. Involvement of local farmers, and impact indicators sensitive to sustainability and poverty criteria, are needed.
- Collaboration with NGOs and CBOs – much collaboration to date has been ad hoc and short term; there needs to be a realisation that the involvement of local NGOs and farmer associations in extension can be effective and long term, and thus worthy of strategic development.

There is scope for exploring the possibilities of more contracting-out of extension tasks to local NGOs or farmer associations. However, NGOs cover a wide variety of competencies and many of their programmes have little independent or systematised evaluation. Some that appear successful at a small scale are too expensive to be scaled up or replicated.

- Collaboration with the commercial sector – there are both opportunities and dangers in this collaboration; guidelines need to be drawn up to prevent bias. Public-funded extension must be orientated to fill gaps left by the commercial sector.
- Divestment of 'non-extension' tasks – divestment of some of these tasks can provide opportunities for redefining core extension roles and creating an optimum allocation of responsibilities between a range of government, commercial and non-profit organisations.
- A more fundamental shift of emphasis is needed within formal training institutions, coupled with improved in-service training in order to produce people with the right skills and attitudes to effectively support sustainable smallholder farming.

Chapter 11

Making Markets Work for Smallholder Sustainability

INTRODUCTION

Successful sustainable intensification will need:

* more use of low external-input, resource-conserving technologies;
* more use of bought inputs – particularly inorganic fertilizer.

Improved input supply can contribute to increased sustainability, as long as its development and promotion does not detract from affirmative action to develop low input techniques.

Price, availability and lack of seasonal credit are all constraints to the increased use of bought inputs. Smallholders in much of Southern Africa pay more for fertilizer (in relation to the price they receive for their produce) than many farmers in Asia – largely due to high transaction costs and market failure. Improving the way input markets work for smallholders can therefore be a major stimulus to sustainable intensification. Improved farm produce markets, giving farmers improved or more reliable returns for their efforts, are equally important to:

* encourage investment in farming;
* enable more inputs to be used;
* where appropriate, fund diversification out of agriculture.

Until recently, most governments controlled the marketing of key agricultural produce using a variety of mechanisms; the purported reasons for this control were a combination of:

- ensuring farmers got fair prices (often pan-territorial);
- ensuring (urban) consumers got cheap food;
- maintenance of food security and emergency grain reserves;
- taxation of agricultural produce.

In fact, the system was often inefficient; expensive to run (marketing boards made large losses); encouraged corruption and patronage; and created policy distortions that were often disadvantageous to agriculture (Box 3.1 illustrates this for Zambia). The liberalisation process in some countries (eg Zimbabwe, Malawi and Zambia) is linked to a structural adjustment programme which is partly dictated by outside actors such as the International Monetary Fund (IMF) and the World Bank.

Liberalisation has transformed the economic environment for smallholders in Southern Africa during the last decade. Details vary from country to country, but in general:

- Governments are withdrawing from the supply and subsidy of inputs.
- Governments are withdrawing from marketing and the control of agricultural prices (privatisation or cost-covering policies for marketing boards and parastatals, ending pan-territorial and pan-seasonal pricing).
- Governments are withdrawing from subsidised, state-run credit schemes.

The process of withdrawal from state-controlled marketing has varied:

- Ending single-channel marketing for various products – this has allowed private marketing alongside the parastatal, which has often been maintained (for a while) to support a floor price and a service in remote areas where there is less interest by commercial companies (eg ADMARC in Malawi, and GMB in Zimbabwe).
- Making the parastatal operate increasingly on commercial principles (ending subsidies) and sometimes privatising the parastatal – this has resulted in closure of depots in more remote or less high volume areas (eg Grain Marketing Board depots in Zimbabwe) and an end to pan-territorial and pan-seasonal pricing.
- Ending the residual buying function of parastatals – this has consequences for the strategic grain reserves in many countries.

This chapter does not dwell too much on the pros and cons of liberalisation since the process is underway and unlikely to be abandoned in the near future; rather, it looks at some of the consequences for smallholder sustainability and the scope for specific actions to encourage more sustainable smallholder farming within the new environment.

CONSEQUENCES OF LIBERALISATION FOR SMALLHOLDER SUSTAINABILITY

Some of the consequences of liberalisation are:

- Reduction in some biases favouring external inputs, such as inorganic fertilizer, caused by earlier subsidies (see Chapter 4) – the ending of these input subsidies has caused a predictable crisis for many small-holders, as bought inputs have become too expensive, yet alternatives to them are underdeveloped. In southern Malawi, with acute land shortage, this is reinforcing a vicious circle of poverty and low yields.
- Input supplies are erratic, with competition tending to be effective in high-use areas with good infrastructure, but not working in many other areas. Thus some areas have a reasonable supply at reasonable prices, and others do not. In Malawi, for instance, in isolated areas of low fertilizer use, the parastatal ADMARC remains the only fertilizer supplier and is often out of stock.
- Smallholders have suffered from increased uncertainty, not knowing what price they are likely to get for their crop. In Karoi, Zimbabwe, smallholders have seen prices rise to over double that of the Grain Marketing Board in the poor year of 1995, when they had little to sell; however, in the good year of 1996, when they had a surplus, the private traders were either not buying, or were offering less than the GMB (MacGarry, pers comm). Such uncertainty makes planning ahead extremely difficult for smallholders with little cash reserves who need to decide whether to take out a loan at the beginning of the season in order to buy inputs.
- Those areas remote from markets, with poor transport infrastructure or with low volumes of production, have experienced either an absolute lack of buyers (eg when the parastatal depot has been closed and no private trader has taken its place), monopolistic marketing with a single or small number of linked traders, or poor prices. While, in some cases, this may stimulate a necessary evolution of production according to comparative advantage, smallholders may need to be supported to make this transition. In others cases, where the problem is lack of current capacity of traders or current poor infrastructure, specific support or investment may be needed to help build the capacity.
- Deregulation of the cereal market in Zimbabwe seems to have had a disproportionately negative effect on small grains, with GMB purchases dropping to a negligible quantity, and with small grain prices not even announced for 1996/97. Some areas managed to sell to Chibuku Breweries, but in many other areas there was no market. Therefore, deregulation may further increase the move to maize, reducing smallholder food security in a drought.

- Although some parastatals, like ADMARC in Malawi, have been expected to retain a role as 'buyer of last resort', and therefore maintain the floor price, this has not always been achieved. ADMARC has at times not had the liquidity to buy from farmers, meaning that actual prices have fallen below the official minimum price. In Zimbabwe, the GMB floor price has become more of a ceiling price, with transporters buying maize at considerably lower prices in communal areas, which they then sell to the GMB.
- While most countries have a strategic grain reserve policy, the parastatals have either not had the liquidity to achieve this, or the prices they were offering have been too low to attract farmers. For instance, the Botswana Agricultural Marketing Board is supposed to keep 80,000 mt in reserve, but in November 1996 it stood at 5000 metric tonnes. Although it is increasingly being argued in the region that national reserves are inefficient, there are dangers if they are completely abandoned (see Chapter 12).
- Market liberalisation has resulted in increased differentials between cereal sale price (farmgate) received by farmers and cereal (or maize meal) price charged by retailers. This is particularly disadvantageous to poorer farmers, who sell grain soon after harvest to meet urgent cash needs (to pay off debts, etc.) and then purchase grain or maize meal later in the year. In Malawi, the maize purchased later in the year may cost twice as much per kilo as the farmer first sold it for.

An additional factor which is likely to have an impact on national markets is the increasing globalisation of markets, linked to an array of trade agreements and organisations (eg the South African Customs Union, SACU; the Southern African Development Community, SADC; Lomé IV; the World Trade Organisation, WTO). These are increasingly restricting the ability of national governments to intervene in agricultural produce marketing, including restricting the imposition of tariffs on exports and imports. While it is beyond the scope of this book to analyse the likely impact of these changes, it is important to note that some of the greatest concern is in the drier countries such as Namibia and Botswana, where production costs are relatively high. In these areas, local production for sale is likely to be undermined by cheaper imports; production for subsistence is likely to continue for those households who do not have other sources of income to buy food; however, these households may find it uneconomic to sell any surplus that they produce.

At a subregional level, free trade is likely to lead to crop production being concentrated in areas with the most suitable natural resources. In southern Malawi, where fertilizer:maize price ratios make the use of inorganic fertilizer uneconomic and small farm areas mean maize tends to be cropped continuously, production is likely to fall. A snapshot of maize prices (see Box 11.1) indicates that, although prices may rise to make fertilizer more economic, an alternative scenario may be increased imports from neighbouring Mozambique, where land is more plentiful and the soil

---

BOX 11.1 PRICES FOR MAIZE IN AND NEAR MALAWI IN DECEMBER 1996

	Kwacha/kg
Reported informal price in Mozambique	0.5–1
Malawi smallholders selling to traders	0.8–1.2
Malawi smallholders selling to ADMARC*	1.55
ADMARC retail sales	2.15
Zimbabwe maize delivered to Lilongwe	3.50
Kenyan maize delivered to Lilongwe	4.00
Overseas maize delivered to Lilongwe	5–5.50

* Parastatal marketing board
Source: EDC, 1997b

---

not (yet) exhausted. This was observed in mid 1997 with Malawians increasingly crossing the border unofficially, both to buy maize and to open fields in Mozambique. Long term sustainability may require more interaction between southern Malawi and Mozambique – and perhaps even more migration from Malawi to Mozambique.

MAKING MARKETS WORK FOR SMALLHOLDER SUSTAINABILITY

There are a number of actions that can be taken to make the liberalised market work better for remote and resource-poor smallholders (see Table 11.1).

Interventions to lower transaction costs are one way of trying to enable farmers to use inorganic fertilizer and other inputs profitably. Transaction costs can be lowered by farmer organisation (see Box 11.2) or by facilitating the establishment of a more widespread and competitive retail network (see Box 11.3).

In Zimbabwe there have also been a number of attempts to build up the local network of retailers, the most ambitious of which is the AGENT scheme (see Box 11.3). The agent programme is interesting – although it has a solid commercial base, there is also another side, with agents being selected by their community with certain expectations for them to provide services, such as advice to farmers. It remains to be seen if the agents on graduation develop differently from typical communal area retailers.

A point to note about the agent programme is the way agent retailers work with agricultural extension and input suppliers to run demonstration plots on the use of inputs. While there is nothing wrong with such demonstrations in themselves, they do contribute to the bias against low external-input alternatives. Affirmative action is needed by extension to make sure alternative, low input techniques are also demonstrated and promoted.

Table 11.1: *Interventions to Reduce Market Failure for Smallholders*

Intervention	Experience and Comments
Lowering transaction costs through farmer organisation.	Farmers' organisations can link to outside service providers or become service providers in their own right (see Chapter 7 and Box 7.2). In South Africa, it is suggested that small new coops in smallholder areas might link with established coops in commercial areas.
Incentives to commercial marketing in remote areas.	For instance: • capital grants to open depots; • organisation and provision of plots (as in growth point development in Zimbabwe); • facilitation of entry into retail business (see Box 11.3); • seasonal credit to grain purchasers and input suppliers; • market days in which a market might be held in a community once a week, attracting a wide variety of mobile traders – these are a way of lowering transaction costs and are traditional in Europe and much of West and Eastern, but not Southern, Africa.
Improved infrastructure.	Lack of feeder roads, bridges and telephones is repeatedly cited as a limiting factor. A report from a multicomponent NGO project in Malawi (see Box 6.2) concluded village access infrastructure was their most successful intervention.
Supporting contract culture (enabling agricultural marketing and rural business to be carried out within a secure, just and legal environment).	By legislation (eg Zambian Agricultural Credit Act 1995); by example – ending the political interference which has contributed to the failure of government credit schemes and undermined the repayment ethos. Levels of theft and violence in some rural areas make production and trading uneconomic.
Market regulation.	Some regulation is needed: • quality regulation; • food hygiene; • pesticides need regulation on human safety and environmental grounds; • accepted standards for organic produce are needed to develop this specific market. However, much regulation in the past has discriminated against small-scale farmers and traders who are unable to meet what are sometimes unnecessarily strict or expensive requirements. Regulations need to be reviewed with smallholder sustainability in mind.
Market intelligence.	Informing small farmers of prices charged or offered can empower farmers and confront monopsonistic suppliers and buyers. A simple example from Mozambique is a noticeboard, run by an NGO, where a long dirt road meets the tar, giving vegetable and other prices in the city in one direction compared with the other direction. On another

scale, the Ministry of Agriculture in Namibia runs a millet marketing intelligence unit and ZFU provides market information and contacts (see Box 7.3).

Improved financial services.	Savings, seasonal credit and insurance are needed to: • enable rational farm planning; • enable seasonal and long term investment; • reduce risk.
Alternative 'temporary' non-profit supply and marketing interventions.	These are quite widespread and are typically run by NGOs or extension offices (eg selling seed, fertilizer, implements); they can be very important in the short term – the difficulty is that interventions like these tend to stifle the development of more sustainable alternatives.

BOX 11.2 MAFAMBIRI FARMERS' CLUB, BUHERA DISTRICT, ZIMBABWE

This is a club, made up of more than 50 relatives and neighbours in a small area of Buhera district, which has been in operation informally for several years, with the objective of providing cheaper seed and fertilizer to members. Around August, the members meet and register the quantity of seed and fertilizer each wants; they agree a date on which payments will be made. On the appointed day, they come together and collect the money; they then appoint one person to go to Harare to make the purchases. This person hires a truck either locally or in Harare to bring back the seed and fertilizer. In previous years, they have collected about Z$22,000 (US$1600) and brought back over 200 bags of fertilizer.

In 1996–1997, they decided against the scheme since prices of fertilizer were too high and the prices received from GMB were too low. Instead, the group managed to get a loan from the district council, which they used to buy fertilizer and to sell it locally. The prices obtained were:

	Harare factory gate (Z$)	Group selling price (Z$)	Local shops (Z$)
Ammonium nitrate	125	160	185–195
Compound D	106	145	165–175

In 1997/98, the group started meeting to plan the purchase of fertilizer and seed; however, then it heard through Agritex that vouchers would be provided for the local purchase of inputs. The group therefore decided not to purchase from Harare, but as of early December was still waiting for the Agritex vouchers.

This group illustrates a number of points:

• the ability of a closely knit group of farmers, with little outside support, to plan ahead and cut the costs of purchased inputs;
• the fact that farmers are sensitive to input and output prices when deciding whether to purchase inputs;
• the interrelation, and perhaps sometimes confusion, between such groups and outside bodies such as Agritex and the council.

Source: EDC, 1998a

Box 11.3 Agribusiness Entrepreneur Network and Training Development Programme (AGENT)

The AGENT programme is implemented by Care International with support from a variety of donors. The objective is to develop agricultural input retailers in communal areas. The programme started in Masvingo and the Midlands in November 1995 with 17 agents, expanded to 60 by March 1997, reached 119 in November 1997, and is planned to reach 500 by 1999 and to include Mashonaland central and east and Manicaland.

Care, along with the community and government staff, identifies potential agents in communal areas; these may be existing grocery retailers, master farmers or ex-teachers – criteria for potential agents are tightly defined. Care provides basic training to the agents, which lasts two to three days and covers information about the agent programme, basic business training and basic information on agricultural inputs and their role in agricultural production; follow-up training and support are also provided, particularly during the first three months. Programme staff, agribusiness suppliers and Agritex are involved in providing training input.

Care acts as the supplier to the agents, receiving orders monthly, combining these and then making bulk orders to the suppliers. In some cases, Care orders direct from manufacturers; in others, Care works through large wholesalers. In one area, Care is involved in setting up a regional wholesaler. Care organises the transport to the agents. Goods are supplied to the agents on a 30-day credit from Care, after which time 3 per cent per month interest is charged. The system is not directly subsidised, with Care charging for transport and levying a 2.5 per cent administration fee. A credit limit of Z\$30,000–50,000 (US\$2,000–4000) is allowed per agent. There is a systematic monitoring and auditing process by Care; unsatisfactory agents receive closer support and monitoring, and may eventually be excluded from the programme.

Typical stock sold by agents is fertilizer (Compound D base dressing and ammonium nitrate), hybrid seeds (mainly maize), insecticides, animal feeds, animal treatments, cement, fencing and doors, and farm implements. The largest sales are of fertilizers.

In the 1996 farming season, the programme initiated 59 demonstrations of maize, sunflower and sorghum, grown to Agritex guidelines and with inputs provided by different suppliers. The stated aims of the demonstrations were:

- overcoming some farmers' perceptions that fertilizer ruins the soil;
- testing a range of crop varieties in dryland conditions;
- presenting a range of alternatives to agents and farmers;
- helping define what works for a particular area.

A pilot scheme is being developed in which the agents will offer marketing services to farmers. Other services may be developed by agents, such as bagging, weighing, storing, and organising of transport for marketing, probably against a small fee. Care has not encouraged the agents to give seasonal credit to farmers, although some agents do this to trusted customers if their liquidity allows it.

Agents are graduated from the scheme, generally after two growing seasons, after which time they are expected to stand on their own feet and deal with suppliers directly, without a credit guarantee from Care. Most of the few agents who have graduated so far are reported to be doing satisfactorily. Care has been trying to encourage the agents to form an association. The scheme was evaluated in mid 1997 (IFAD, 1997) and it was found that:

- The pilot programme had demonstrated that the approach to improve input supply, based on market forces, was viable.
- There might be scope for future broadening of the approach to include livestock and crop marketing and food processing.
- Agent turnover averaged US$1000 per month and the initial 17 agents received US$220,000 worth of inputs, with 95 per cent of all agent payments being made on time and no agent yet defaulting.
- Agents were reaching 55 per cent of farmers (an average of 1100 farmers out of 2000) in their catchment area, with farmers having to travel an average of five kilometres. Farmers make an average of five visits to the agent per year and include purchases of both seed and fertilizer. Farmer yearly purchases were in the range Z$500–3000 (US$30–200). On average, agents were sited 16 kilometres apart.
- The mark-up of agents ranged from 5 to 30 per cent with an average of 16 per cent; most thought the mark-up would increase next year, but this would depend on competition.
- Most farmers who purchased inputs suggest that these are incremental purchases (rather than a simple transfer of retail outlet use).
- It was considered too early to assess sustainability of the programme. The possibilities for scaling up were considered to be very good; it was noted that there was considerable interest in replicating the programme further within the region, including the possibility of franchising the approach.
- The demonstrations had shown potential successful links between input supply and extension efforts. The demonstrations had been mainly concerned with hybrid maize and fertilizer.
- Of the 16 agents surveyed, 11 had made some credit sales to customers, and half of these had experienced difficulties in obtaining repayments. The evaluators felt increased provision of credit by agents would endanger the sustainability of the programme.

Source: EDC, 1998a

With the liberalisation of seed markets, commercial companies have increasingly been entering national markets, often in competition with existing parastatal companies. Although increased competition may bring new opportunities for smallholders, it is also important that some positive aspects of the previous system are not lost. Commercial companies have understandably been most interested in selling hybrid varieties of the major crops, particularly maize. There is much less interest in open pollinated varieties and minor crops, which can, however, be very important for smallholder diversification and risk management. Some specific interventions, trying to combine commercial principles with a public good input, may be needed to maintain access to suitable seed. Box 11.4 gives an example of the Namibian government actively encouraging a small-

---

Box 11.4 Northern Namibia Farmers Seedgrowers' Cooperative

In northern Namibia, millet is still the established staple, and although there is a wide variety of maize and brewery sorghum seeds on the commercial market, farmers growing pearl millet, food sorghum, cowpeas and bambaranuts have until recently been restricted to local landraces. While these are adapted to varying local soil potential and rainfall, they are mainly late maturing. In response to farmer demand, an early maturing variety of millet has gained rapid popularity among Namibia's smallholder farmers, although, wisely, most farmers seem keen to grow a mixture of varieties, including traditional landraces, both to spread risk and for reasons of taste. Developing a sustainable system of seed supply of the early maturing, open-pollinated variety has been one component of the process.

The introduction in the 1989/90 season by a local NGO, the Rossing Foundation, of an early maturing millet variety – Okashana 1 – after some trials carried out in cooperation with ICRISAT, resulted in rapid uptake by farmers in many areas. Having produced and released a new crop cultivar, the next issue was how to provide a sustainable supply of seed.

Seed bulking and distribution were at first undertaken by the government. An attempt is now being made to hand seed bulking over to the private sector in the form of a government-sponsored cooperative, comprising small-scale seed growers in the Omusati region. Using a number of small-scale seed growers spreads the risk of seed crop failure and gives these farmers a profitable cash crop. Farmers must dedicate two to five hectares to seed production; in order to avoid contamination, seed production fields must be at least 200 metres from other pearl millet fields, and in areas not infested with wild-type millets. Inspection of participating farmers' fields is carried out by the extension service.

In the 1995/96 season, 230 tons of seed were produced (213 tons by 93 farmers, and 17 by the ministry). While demand from within Namibia is estimated to be about 150 tons annually, other potential markets include Botswana, Malawi, Zimbabwe, Angola, South Africa, and even Sudan. In future, sorghum and cowpea seed may be produced by the cooperative.

Source: EDC, 1997d

---

holder farmer cooperative in producing seed; Box 11.5 is an example of an NGO, commercial company and farmers linking to do the same thing in Zimbabwe.

FINANCIAL SERVICES FOR SUSTAINABLE AGRICULTURE

Financial services include both formal and informal institutions, providing mechanisms for saving, credit and insurance. These services are important for smallholder sustainability in a number of ways:

- They facilitate rational farm management by acting as a buffer between a household's fluctuating cash needs (school fees, medical

Box 11.5 ENDA Seed Action Project, Zimbabwe

A number of organisations, including Environment and Development Action – Zimbabwe (ENDA–ZW), are promoting small grains and other drought-resistant crops in Zimbabwe. This is because it is felt that small grains are more appropriate in drier areas, and that the seed companies, Agritex and GMB, have tended to promote maize at the expense of small grains.

Initially, a range of organisations worked together to make small grain landrace collections, which were then evaluated and stored. ICRISAT helped with breeding and introducing germplasm from outside. However, further action was needed to make small grain seed available on the commercial market.

- ENDA contracted smallholder farmers to grow small grain seeds; by 1997 they were working with 3000 growers, much of the product marketed through the large national seed company SeedCo. Currently ENDA is investigating setting up an independent company for marketing small grain seeds to be owned jointly by SeedCo, the farmers and ENDA.
- A dehuller was developed to work in tandem with local grinding mills, making them able to mill small grains. This dehuller is now in commercial production, independent of ENDA.
- Lobbying is used to raise the profile of small grains; some of this is being done in conjunction with other organisations coordinated through the Zimbabwe Seeds Action Network. Small grain seeds have now been included in government drought packs. They have also lobbied to relax some of the seed regulations which are considered unnecessarily restrictive.
- Small grains still face unfavourable price differentials, with the marketing board paying Z$700 (US$45) per tonne for sorghum, compared with Z$1200 (US$80) for maize. Shortly after independence, when the price was more favourable towards small grains, the marketing board bought considerable quantities of small grain that it was unable to sell. ENDA is therefore working with other organisations trying to improve the market for small grains – for instance, as ground meal, or as stock feed. Currently, those farmers able to market to brewing companies receive relatively favourable prices.
- There are some constraints at farm level – small grains can be more susceptible to bird damage and may require more labour than maize.

Lessons learned

- Action on a number of different fronts is needed to overcome a series of constraints such as germplasm collection; seed production and sales; bird damage and labour demand; milling technology; advocacy; and marketing. Progress on marketing and farmer level constraints continue to be limiting.
- Networking with other organisations can be necessary to produce the required impact, although it has been difficult to maintain the required intensity.
- Despite the involvement of a range of NGOs, ICRISAT and SeedCo, and an active advocacy component, the project has barely broken into the mainstream. The bias in favour of maize remains strong.
- It is necessary to look again at some of the reasons why many farmers continue to shift to maize; birds, in particular, can be a problem, especially where only a minority of farmers are growing small grains so that bird damage is concentrated on the few fields with small grains.
- In addition to the commercial production of small grain seed, there is probably scope for more support for on-farm small-grain seed selection and storage.

Source: EDC 1998a

costs, etc) and optimum income-generating strategy – livestock can be sold when they are ready or prices favourable, rather than the moment the cash is needed.

- They can provide the cash needed, on either a seasonal or longer term basis, to invest in increased production (eg buying fertilizer, implements or livestock) or indeed to invest in resource-conserving technologies (such as fallows and tree crops).
- By reducing risk through having savings, and by diversifying risk between livestock and a saving institution, or through informal or formal insurance, they enable longer term planning and investment in sustainability.

There has been generally more emphasis on the credit needs of small-holders in Southern Africa than on savings or insurance. However, there is a growing realisation that savings are equally important. Subsidised credit schemes, designed to encourage seasonal inputs, have been part of the bias towards bought external inputs which was described in Chapter 4. Experience in formal sector smallholder insurance is underdeveloped, although there is an insurance element in some traditional savings groups and burial societies (see Box 11.8).

There has also been greater concentration on formal financial services, with often limited understanding of community-based mechanisms operating between relatives and neighbours. There is recent interest in supporting community systems and in some cases trying to link these to formal sector institutions.

Financial services in rural Southern Africa tend to suffer from high transaction costs:

- Information is costly where communication is poor, populations are dispersed and the external economic situation is in flux.
- There is often no reliable legal mechanism for enforcement of contracts and many smallholders do not have collateral for loans.
- Group systems for providing mutual liability can be expensive to nurture and to sustain.
- Climatic variability increases risk for all involved.

Many of the countries in the region have seen a catastrophic collapse of their government-run credit schemes during the last decade. Although these schemes were often inefficient, subject to political patronage and tended to reach the richer farmers, they were still quite significant – reaching 30 per cent of smallholders in Malawi and 10 per cent in Zimbabwe. Rebuilding these credit schemes is proving difficult (see Box 11.6); there are moves in most countries to do this on more commercial and sustainable lines. The question is whether the more commercial approach will include the poorest smallholders.

With difficulties experienced by national credit programmes, and the recognition that much of their credit is targeted to the better-off small-

Box 11.6 Rebuilding Smallholder Credit at the National Level

Malawi

Before 1992, the parastatal smallholder credit agency, SACA of Malawi, managed a system in which there was close interlinking of the group credit approach, agricultural extension, the Malawi Congress party, and the promotion of subsidised hybrid maize and fertilizer. This system was highly regarded internationally because of its high repayment rates, although it only reached about 30 per cent of the richer smallholders and benefited from the subsidy of the free input of the extension service. Extension workers managed the credit groups, and farmers had to attend group meetings and follow the extension workers' recommendations in order to get credit, which was typically seasonal credit for hybrid maize seed and fertilizer. Coercion was occasionally used, often linked to the party youth structure, to ensure repayment of loans. The system foundered, due to a number of factors:

- The system expanded beyond its management capacity.
- Political changes reduced repayment discipline and the scope to use coercion.
- The use of fertilizer on maize became unprofitable once subsidies were withdrawn, prices liberalised and the Kwacha devalued (Chapter 4).

The number of borrowers dropped from 388,000 in 1992 to 45,000 in 1993; SACA collapsed in 1994. SACA has been replaced by the Malawi Rural Finance Company, which is supposed to be more independent of political influence and to follow market principles. Consequently, it has imposed market interest rates (44 per cent in 1996/7), eschewed lending for maize, and restricted its lending to profitable sectors such as tobacco. The exception is in the Tikolore Clubs (see Box 11.7), for which it receives a specific subsidy to cover higher initial transaction costs, although the actual loans and interest are supposed to be made on a commercial basis.

Zambia

Before liberalisation, credit for maize input supplies to smallholders in Zambia was supplied through the cooperative movement (although with considerable government involvement). Although some aspects of the organisation were weak, this had the advantage of being able to sustain relatively high rates of recovery through a monopolistic tie-up with marketing (World Bank, 1992: 56).

With the liberalisation of the fertilizer and maize trade, responsibility for crop finance was transferred to the Credit Union and Savings Association, ZCF Financial Services and Lima Bank. Without monopoly control over marketing, however, their organisational weaknesses and financial dependence upon government were exposed (Lekgetha et Al, 1995; GRZ/MAFF, 1991). Drought and late delivery of inputs were additional factors behind poor recovery. The problem was compounded and confused by direct political intervention in credit operations, with politicians encouraging farmers to believe that they would be able to secure credit in the future, even if they failed to repay. All three institutions currently have little capital for new lending, and while the government has dithered over how much outstanding credit from the 1994/95 drought year to write off, the banks themselves have attracted adverse publicity by seizing collateral (eg *Zambian Farmer*, December 1995: 15 & February 1996: 1).

In 1995/96, the Zambian government launched the Agricultural Credit Management Programme. During the first two years of operation, the management of fertilizer credit has been subcontracted to private firms (Cavmont Merchant Bank and SGS Ltd), who have in turn hired small traders to act as input distributors, marketing and loan collection agents. The programme got off to a bad start as a result of political interference in the selection of local agents (*Zambian Farmer*, January 1996: 2), and repayment rates have been very low, with many farmers continuing to regard the fertilizer distributed under the credit programme as a government hand-out, despite the private sector intermediation (GRZ/FAO, 1996b).

The problem of supplying seasonal agricultural credit in a sustainable way has clearly not yet been resolved. It is clear that simple contracting out of the management of credit to the private sector is not a solution, if the underlying problems are not also solved.

Zimbabwe

Seasonal loans by Agricultural Finance Corporation (AFC) to communal area farmers increased rapidly during the 1980s, up to over 70,000 in 1986. However, a combination of high transaction costs, low repayment rates and wider economic policy changes, with the government being increasingly unwilling to underwrite losses by the AFC, led in the late 1980s to a rapid fall in the number of loans to under 30,000 in 1991. The AFC, although continuing to lend to CA farmers, directly reaches less than 3 per cent of farmers; these tend to be the better-resourced farmers, often for cash crops and with collateral to provide security.

The AFC has continued to be pressurised by government to provide credit to smallholders. In the 1990s, it tried to provide credit to communal area agriculture in ways that overcame some of the high transaction costs and low repayments experienced in the 1980s:

- **Lending through intermediaries:** this is a new programme started during the last two years; an example is the SHDF (see Box 11.8). The advantages are that the intermediary may have lower transaction costs, may be better placed to assess credit worthiness of borrowers and can enforce repayment.
- **Group lending:** this started in 1992 and currently 40 per cent of communal area lending is to groups. Groups have joint liability for their members' loans, and repayment rates are reported to be higher than that of individual smallholder loans. Attempts to keep transaction costs low mean the AFC insists groups are fairly large, often resulting in groups which lack cohesion; the ZFU and others feel that some groups are too large to be manageable.
- **Input suppliers:** the AFC lends to one supplier per growth point, enabling the supplier to hold larger stocks of inputs; this component has been running for four years.
- **Grain buyers:** this is a new programme under consideration for 1998.

While these new programmes are probably realistic and more sustainable ways for a large organisation such as AFC to support CA agriculture in a market environment, it is not certain that these programmes will be developed and sustained. Future policy seems to be moving towards greater independence from government, which will probably mean more emphasis on commercial and less on developmental activities, with likely rises in interest rates and a reduction in the proportion of loans to small borrowers.

Namibia

Namibia has not had to deal with a history of a failed scheme and has recently launched a National Agricultural Credit Programme through the parastatal Agribank, which previously serviced the large-scale sector. Some of the features of this programme are similar to earlier state-sponsored schemes in the region which have failed (interest rates are subsidised, there are links to the extension service and loans seem mainly to be going to richer smallholders). It is still too early to say whether the lessons from elsewhere in the region have been learned and whether the scheme will be sustainable, and how it will affect poorer farmers.

Source: EDC, 1997b; 1997c; 1997d; 1998a

holder, there is interest in developing ways to reach poorer farmers. One attempt at creating a financially disciplined scheme that is specifically targeted at poorer women smallholders is the setting up of the Tikolore clubs in Malawi (see Box 11.7).

There are a range of smaller-scale initiatives on credit, often NGO led. Some NGO programmes have been criticised in the past for charging uneconomically low rates of interest, and therefore stifling the development of more sustainable alternatives (EDC, 1997b); however, it is a rapidly developing sector, with increasing networking between organisations and increasing recognition that, for sustainability, programmes must be able to cover their costs, which usually means that interest rates approach commercial rates. The challenge is to cover costs, while still serving the poorest households.

Rural Traders – a Sustainable Alternative to Group Credit?

Much of the emphasis on credit development in the region has been on group schemes, using the group to lower transaction costs and to provide a degree of mutual liability through knowledge and ongoing relationships with each other. An alternative route can be through rural trading stores – which already have a transaction infrastructure and already know many of their customers. In the past, well-established rural trading stores were sometimes an important source of credit to smallholders; with economic liberalisation, these stores are being reestablished or are expanding into agricultural supply and marketing. Many, however, face severe liquidity problems and have difficulty accessing seasonal credit for themselves, let alone to pass on to their customers. It is possible that improved credit access for these traders may, in turn, improve their ability to give credit to smallholders (an example is the AFC programme of lending to traders in Zimbabwe – see Box 11.6).

Trader-mediated credit may impose adverse conditions on small farmers (eg high interest rates or only lending to the rich); this is likely to

Box 11.7 Malawi's Tikolore Food Security Clubs

Malawi has witnessed several efforts to assist poor families with credit packages to help them overcome chronic food shortages. None have succeeded because the farmers (mostly women) have had to sell maize in order to repay their loans. In most cases, the incremental maize which they produced as a result of the credit (usually to buy fertilizer and hybrid seed) was insufficient to enable them to pay off their debts, feed their families and save enough to purchase inputs for the next year, as was intended by the programmes (mainly due to the unfavourable fertilizer:maize price ratio – see Chapter 4). In many cases, the women had to make a choice between feeding their children and repaying their loans. The former naturally took precedence, and so the schemes were short lived.

On the basis of this experience, the Malawi government proposed an alternative strategy, with financial backing from IFAD for its implementation. As in the past, this new food security project is based on providing credit to poor families, with emphasis on women. The fundamental difference is that the family is issued with credit for a small area of cash crop (tobacco) in addition to a loan for seed and fertilizer for food crops (maize and soya beans). The proceeds from the high-value cash crops are used to pay off the total loan, so that all the food stays in the family.

Credit is issued through a special window of the parastatal Malawi Rural Finance Company to these Tikolore ('let us gather/harvest') clubs. Farmers are grouped into clubs of five members and five such clubs form a 'centre', which is the administrative basis of the scheme. The five members of a club are responsible for helping each other, particularly in sickness, and are jointly liable for repaying their loan. Their tobacco is sold in the name of the centre and the loan is deducted at source for the whole group and not on an individual basis. Club members have to work out their repayment situation at the local level. In a good season, there should be a surplus of actual money from the sale of the tobacco, after all the expenses have been paid, but the main objective of the programme is that the family should grow some 300 kilogrammes of extra food in the home, with no need for any of it to be sold to repay debts for inputs. This should be sufficient to carry a poor family through the three hungry months prior to the annual harvest.

Two features of the clubs make them self-selecting for poor families. The first is the requirement that all participants attend ten weekly training sessions on credit, and the concept of working as groups. At the end of this training, they have to pass an examination to evaluate their understanding of the Tikolore concept. The second is that loans which they receive after this training are modest, amounting to approximately UK£30 per family. Wealthier families are less interested in loans of this size and would be unlikely to attend ten classes in order to obtain such a sum. The combination of small loans, plus the classes, tends to result in participation by the intended beneficiaries, without any invidious inquiries into people's poverty.

The loans are made at the standard rate of credit company interest (currently 44 per cent per annum), but it is recognised that the screening and training of these groups incurs the company a good deal more expenditure than loans to its wealthier clients. This has been especially true in the first year, when training manuals have had to be prepared, staff courses organised and extra-close supervision provided. The company has been given a grant of US$100,000 per year by IFAD for running the scheme.

Small pilot programmes were run to test the scheme and these demonstrated that, contrary to many people's expectations, poorer women were capable of producing and marketing a cash crop and handling their financial affairs. On the basis of this experience, the scheme started on its full programme in 1996 with some 600 clubs. This number is expected to grow to over 3000 clubs during the next two to three years.

Source: EDC, 1997b

BOX 11.8 SELF-HELP DEVELOPMENT FOUNDATION (SHDF)

The Self-Help Development Foundation is one of Zimbabwe's oldest and largest local NGOs, although it is not particularly well known. SHDF began in 1963 as a way of supporting saving by rural people with the motto: 'take care of your money today, so that it will take care of you tomorrow'.

Saving takes place in clubs; these typically have about 25 members. Clubs are a means of providing peer group support and discipline, and a means of overcoming distances to commercial savings institutions and the reluctance of these institutions to deal with small deposits and withdrawals (small for the bank, not for the rural saver). Club members contribute on a regular basis, perhaps Z$10 (US$0.70) per month. The total contributions are then taken by one member and deposited in a saving institution such as the post office, with members contributing to transport costs if necessary. Clubs are independent and their rules vary on the contributions required and on conditions for withdrawal. The majority of clubs are for seasonal agricultural inputs, but others also operate to pay school fees or to pay funeral expenses, the latter providing a type of insurance function. Some clubs are organised around specific productive activities such as handicrafts, or horticulture, and may include joint purchase of inputs or joint marketing arrangements.

SHDF's input is mainly involved in promotion and training activities; they have a coordinator and two field officers in each province. There are also volunteer promoters at community level; although not paid a salary, they are sometimes paid for arranging training courses. SHDF provides record books for clubs and some clubs use a system of stamps to record savings. Since clubs are independent, SHDF does not guarantee the money saved by individuals in the clubs; however, theft by treasurers is reported to be very low since a club is typically made up of relatives or people who know each other well. The number of clubs is reported to be expanding and reached an estimated 12,000 in 1997, involving approximately 300,000 people. Over 90 per cent of members are women.

In 1996, SHDF took what it describes as its biggest step since its foundation by moving into supplying credit. Credit is provided to SHDF by the Agricultural Finance Corporation at an annual interest rate of 25 per cent. SHDF then loans this money either to whole clubs or to individuals within a savings club. Credit worthiness is assessed by the savings record of the individual or group and the viability of the project proposed. Loans are for a period of six to 12 months, are in the range of Z$2,000–25,000 (US$130–1,700) and can be up to five times the balance held in the savings club. 5 per cent interest per month is charged on the outstanding balance.

Loans are very popular among savings clubs and it is reported that members do not consider the interest rate excessive. So far, 50 loans have been made, totalling two million Z$ (US$130,000), many of them for agricultural projects. There has been a 100 per cent repayment record so far, with many loans being paid back early. It is estimated that 100 loans will be needed at any time to cover the programme's running costs.

Organisations such as SHDF and a number of others involved in micro finance have recently started the Zimbabwe Association of Micro Finance to exchange experience in what is a rapidly developing field.

Source: EDC, 1998a

be minimsed where there is a reasonable degree of competition between different traders and between traders and group-based lending schemes. Trader credit is also likely to be tied to specific inputs purchased in the trader's store – rather than for unpurchased inputs (such as paying someone to ridge your fields) which may be more resource conserving. However, it is also possible that in future traders could become more general providers of financial services – not just provide inputs on credit.

Linking Savings and Credit

It is increasingly recognised that credit is only part of the story – savings opportunities are also important for smallholders and, worldwide, many of the best credit programmes are linked to savings. There are also indigenous savings and credit traditions in much of the region, for instance *stokvels* in South Africa. There appears to be scope for building on some of these traditions; the SHDF is trying to link grassroots savings groups with credit opportunities (see Box 11.8).

Commercial savings and loan facilities for smallholders tend to be underdeveloped in the region – in Namibia, until recently, only three commercial banks served 60 per cent of the total national population living in the northern communal areas. In Zimbabwe, the credit union movement is supporting the development of village banks as an alternative to commercial facilities (see Box 11.9).

Although village banks, such as the one in Dema, seem to be meeting an important need, there is also a risk in such schemes – this is because savers and borrowers who are largely dependent on agriculture incur similar risks (co-variant risk). Therefore, in a drought, many of the borrowers may not be able to repay and many depositors will also need to withdraw their money – leading to disaster. It is important that small rural banks are able to diversify their risks, usually meaning linkages with an apex organisation, links to commercial banks or through insurance.

For many smallholders, livestock remains the main form of savings. Recent droughts have undermined the security of this investment. Using livestock as a saving is thought to encourage higher stocking rates than those optimum for meat production. There is a theoretical argument that increased savings institutions will reduce the importance of livestock as savings, with knock-on consequences for reduced risk and increased livestock off-take. However, evidence for this actually occurring is difficult to substantiate.

AGRIBUSINESS-LED LINKAGES TO SMALLHOLDERS

An alternative marketing system, which seems to be expanding rapidly in the region, is through agribusiness-led linkages to smallholders such as outgrower schemes (Stringfellow, 1996) or interlocked contracts, where

inputs are supplied on condition that the crop is marketed through the
input supplier; these have recently been reviewed by Poulton et al, 1998.
Typically they involve a tie-up between the provision of inputs and advice
and a contractual obligation to sell to the agribusiness input provider. In
Zambia, cotton outgrower schemes have been stimulated by the sale of ex-
government processing enterprises to the private sector. In Zimbabwe, the
move of commercial farmers out of cotton has stimulated CottCo to
expand purchases from smallholders, and competition between lint
companies in a newly liberalised market seems to be providing some
benefits for growers, in the short term at least (see Box 11.10).

Some of the issues arising from the outgrower expansion, particularly
in cotton are:

- Cotton growing can be a profitable crop for some smallholders and
  has the advantage that it is suitable in a relatively dry region, where
  there are few alternatives. However, those able to benefit are the
  better-resourced farmers, with land and labour, who are able to meet
  minimum (and increasing) production quotas.
- Initial competition between buyers seems to have brought price
  benefits to smallholders. However, as the sector stabilises, this
  competition may be reduced, with oligopolistic relations between
  buyers, group leaders loyal to specific companies, and with producers
  tied into forward contracts.
- The ability of a small number of private extension workers to manage
  the CottCo programme is instructive – this is due to the rigid rules
  and the incentives for group members and leaders to obey the rules
  and stay in the scheme. To what extent discipline will be maintained
  as the range of potential buyers increases is still unclear. The organisational lessons are probably more relevant to managing credit
  schemes than to more generalised extension programmes.
- Cotton and some other cash crops are heavy users of pesticides (this
  is a major reason for the outgrower model), with consequent risks to
  human and environmental health. Companies need to take a more
  active role in promoting agrochemical safety. This is an aspect which
  requires further research and monitoring. Increasing competition
  between companies is unlikely to promote more health and safety
  activities by the companies and is therefore a potential area requiring
  government regulation.
- Cotton production is sensitive to fluctuation in world market prices,
  and the risk of changes in dollar-denominated input/output price
  ratios is currently borne by the outgrower during the season
  (however, it is suggested that a 63 per cent fall in prices would be
  needed in Zambia to wipe out current smallholder margins).
- There is a gender dimension – cotton cultivation may result in
  reduced inputs into food and women's crops, and thus have an
  adverse nutritional effect.

Photograph 11.1: *Mixing pesticides on top of the village water supply –
and without gloves. Will private chemical suppliers and outgrower schemes
provide sufficient training and safety equipment, or will regulation be
needed to enforce this?*

There is also an increase in outgrower-type schemes which produce high
value crops for northern markets, such as out of season vegetables and cut
flowers for European supermarkets. So far, most such schemes involve large-
and medium-scale farmers, but there is potential for more involvement by
smallholders. Such markets, as an additional opportunity for smallholders,
are likely to bring some benefits. There are also potential risks:

- Too great a dependence on any changeable market like this can bring
  increased risks.
- There is a gender dynamic: if traditional food crop land is converted
  to cash crops, with the income being captured by men, there is a
  potential loss to the wider household wellbeing, particularly child
  nutrition.

Box 11.9 Village Banks in Zimbabwe

Credit unions are basically financial cooperatives owned and run by their members; they are supported by the National Association of Credit Unions of Zimbabwe (NASCUZ). The early unions in Zimbabwe were typically organised in a particular workplace, with weekly or monthly contributions made by workers who, in turn, had the right to borrow for emergencies – such as funerals – or for investments – such as starting a business or for agricultural inputs. There are currently 70 registered credit unions, with a membership of about 34,000.

During the last three years, there has been a programme of spreading the credit union movement to rural areas with the development of village banks. A rural credit union or village bank needs about 300 members to be financially viable. Members invest in the village bank by:

- buying shares – non-withdrawable except on resignation;
- regular savings, for instance for school fees;
- ad hoc deposits, which can be withdrawn at will.

Members receive interest on their deposits and can also apply for loans. Typically, the spread between interest on savings and on loans is about five to 12 per cent, which finances the running costs of the bank. There is considerable pressure to provide competitive interest rates in comparison to the post office, building societies and banks. Although post offices are relatively widespread in rural areas and provide interest on savings, they do not give loans. Banks and building societies are confined to urban centers.

Individual credit unions are independent. NASCUZ provides training and auditing facilities, and unions make contributions to NASCUZ to help pay for their services. NASCUZ also provides a central finance facility as part of a guarantee scheme and to help iron out seasonal cash-flow problems (November to December is a critical period, due to the combination of agricultural inputs and Christmas). NASCUZ also raises loans on the money market, which it 'on-loans' to individual unions on a commercial basis.

The village bank programme is relatively new; initial results look promising, but long term sustainability will in part depend on building up sufficient turnover to cover overheads and maintain investor confidence. The bank in Dema is an example of a recently opened bank.

Village Bank, Dema

The village bank at Dema, in Seke district, is run by the Neshungu Savings and Credit Cooperative. The Coop was registered in 1995, although it was in existence before this, and the bank opened its doors in August 1996. In December 1997, it had 290 members, with five to ten new members joining every month; a small majority of the members are women. Members pay a joining fee of Z$30 (US$2), buy from one to ten shares at Z$100 (US$6) each and must deposit a minimum balance of Z$50 (US$3). At least one savings club is a member and some group development projects (eg chicken rearing) are also members.

During the last two years the members have been assisting in building the village bank. NASCUZ obtained donor funds for the building material and for paying the builder, while the members made blocks and provided sand and other assistance. NASCUZ and Silveira House, an NGO which has a long term involvement in savings schemes, provided technical support and training for the manager and the various committees.

The projected balance for the end of December 1997 was (in Zimbabwe $ with approximately 15Z$ = 1 US$):

Liabilities		Assets	
Savings	900,000	Cash in hand	10,000
Loan from NASCUZ	36,000	Bank savings	30,000
		CFF investment	70,000
		Bank investment	80,000
		Loans to members	1,045,000
Capital		Fixed Assets	
Shares	300,000	Building	90,000
Building fund	100,000	Other	10,000
	1,336,000		1,336,000

About five to 20 customers use the bank each day. In addition to serving members, the bank also cashes cheques, for a fee, for non-members. Members receive 11 per cent interest on savings and pay 26 per cent interest on loans. The nearest commercial bank is 30 kilometres away and charges 38 per cent interest on loans.

Clearly, it is still early days for the Dema village bank – crucial to its success will be whether it manages to achieve a high level of repayment on its loans and sufficient turnover to cover its operating costs. A few other village banks have been operating for slightly longer than Dema, and up to now their performance has been good.

Source: EDC,1998a

This is a sector where some Northern consumer groups actively encourage fair trade with Southern producers – such campaigning can bring benefits as long as it is well informed about the Southern smallholder reality. A major drawback of most of this new 'luxury' market is its reliance on flying fresh produce to the North – this contributes to the greenhouse effect and higher level atmospheric pollution and cannot be considered sustainable.

It is not clear whether there will be a continued expansion of outgrower schemes, or whether this is, in fact, a transition, and that as a greater diversity of credit, input and marketing channels develop, a more open (and possibly competitive) input and marketing environment for cash crops will develop. In general, smallholders will benefit from having choices; thus linkage and outgrower schemes are not detrimental in themselves, but can be if they are the only option. Development of alternatives to outgrower schemes may enable smallholders to negotiate from a position of strength on how to participate.

INPUT SUBSIDIES AND CONTINUED GOVERNMENT INTERVENTION

Despite a more general move away from subsidies as part of economic liberalisation in the region, there remain several initiatives which provide subsidies to inputs. There are a number of reasons for such subsidies:

Box 11.10 Rise in Cotton Outgrower Schemes in Zambia and Zimbabwe

Cotton is by far the most important industrial crop for Zambian smallholders. Substantial investments in spun yarn production following privatisation mean that domestic demand is outstripping supply. In response, cotton has been called the 'white emerald' and there has been significant investment in cotton outgrower schemes (particularly by Lonrho, which took over the parastatal Lintco in 1996). The World Bank forecasts that outgrower schemes will increase from an average of 62,000 smallholders for 1990–1995, to 95,000 by the year 2000, with area cultivated per farmer staying constant, but with yields growing by 50 per cent (World Bank, 1996).

The Cotton Company of Zimbabwe (CottCo), formerly the cotton marketing board, first became a parastatal company owned by the government, and since October 1997 it has been a private company, although the government retains a minority holding. Since 1992, the company has been building up a programme of contract growing with smallholders, initially using World Bank credit. The number of participants reached a peak of 86,000 in 1995/96. With the ending of the company's monopoly over cotton purchasing, other players, such as Cargills, have entered the market and the number of CottCo contract growers fell to 75,000 in 1996/97. The number of growers is expected to fall again in 1997/98, but after this the company expects contract discipline will improve and stability will be achieved. In 1996/97, competition between the various cotton buyers caused the price of cotton to rise several times during the year.

The company provides inputs to farmers, organised in groups, in a carefully controlled system aimed at managing risk. The farmer signs a contract agreeing to sell the whole crop to CottCo (just repaying the inputs is not sufficient and any alternative sales result in the farmer being excluded from further contracts). A farmer's three-year delivery record is used to assess probable yield, then inputs are provided on credit to a value of up to 45 per cent of this yield. However, these inputs are provided in three tranches:

- tranche 1 – seed and basal dressing;
- tranche 2 – top dressing and agrochemicals;
- tranche 3 – agrochemicals.

Before tranche 2 and 3 are given out, a crop assessment is made and the credit limit adjusted as necessary.

Farmers are organised on a group basis, with the group collectively liable for defaults. Group leaders are appointed and they assess crops and recommend second and third tranches. The group leaders are paid bus fares, meeting attendance allowances, and a proportion of the total sales of the group (as long as all the group sales go to CottCo; otherwise they get nothing).

The system is managed by 16 CottCo extension officers who deal with 2500 group leaders (approximately 150 each); the extension workers have a pick-up truck or motorbike. There is also close collaboration with Agritex and the ZFU. To maintain cost effectiveness, there is a limit of 400 kilogrammes of cotton, below which a farmer is not allowed to join the scheme. CottCo plans to raise this gradually to about 1000 kilogrammes.

In 1996/97, CottCo sold 5000 sprayers, but it apparently does not sell either overalls or face masks.

Source: EDC,1997c & 1998a

Box 11.11 Namibia – Input Voucher Scheme

As a condition of divestment from other agricultural services (see Box 10.3), the Namibian government agreed to a voucher scheme to the value of about N$70 (US$16) per household, to be redeemed in part payment for commercial ploughing services (tractor or animal drawn), fertilizer and certified seeds. The cost of this voucher scheme was equivalent to the cost to the government of operating the various services that are handed over to the private sector. The voucher scheme is targeted at some 65,000 northern communal crop-growing farming households which are classified as poor because they are amongst the 46 per cent of households spending more than 60 per cent of their total consumption expenditure on food (according to the Namibia Household Income and Expenditure Survey of 1993; CSO, 1996).

Weak management has meant that, at the time of writing, there is virtually no progress in implementing this programme. It has been suggested that as a long term transfer measure, the agricultural input voucher scheme should be designed so that it can be upgraded after drought years to compensate poorer farmers for loss of vital inputs (eg seeds) and other assets (eg draught animals and implements).

Source: EDC,1997d

- As a type of *surrogate social security*, targeted (in theory) on the poorest households: these programmes are perhaps of most immediate interest in countries such as Botswana and Namibia, with relatively high GDP per capita but seemingly intractable rural poverty. The Namibia voucher scheme falls into this category (see Box 11.11).
- As part of a more general *drought recovery programme*: however, there is a tendency for such programmes to be continued, sometimes becoming semi-permanent features, perhaps for reasons of political expediency. In Botswana, the Accelerated Rainfed Agricultural Programme paid for ploughing and destumping etc. It cost a lot of money, tied up considerable extension time and produced little long term impact (EDC, 1997a). In Zimbabwe the post-drought crop pack, with its high transaction costs, has been replaced in 1997/98 by vouchers covering part of input costs.
- Subsidising investment in farm productivity: the ALDEP in Botswana programme is an example (see Box 11.12).

There are also some sectors and countries where government controls seem likely to continue or perhaps expand. Export meat-marketing channels, particularly in Namibia and Botswana, remain controlled, partly because of the strict quarantine and quality controls, linked to export agreements to the EU. In Namibia, the company MeatCo was created by the pre-independence privatisation of state assets; however, political pressure encouraged its expansion into the northern communal areas, operating unitary grading and pricing policies (this is seen as a political necessity in Namibia, in order to give the northern communal area farmers

BOX 11.12 ARABLE LANDS DEVELOPMENT PROGRAMME (ALDEP)

ALDEP has been Botswana's most important programme in support of smallholder dryland agriculture. The main emphasis has been on promoting a technology package by providing highly subsidised inputs to selected farmers. ALDEP started in 1982, financed by IFAD and ADB, running through to 1996. It is currently entering a second phase, to run from 1997 to 2003, entirely funded by the Botswana government.

Originally ALDEP was conceived as a loan and subsidy scheme. However, the transaction costs of servicing large numbers of small loans were not viable (the National Development Bank estimated a cost of P110 (US$35) per year to service a P300 (US$100) loan over five years), and the scheme switched to 85 per cent grant and 15 per cent down-payment by the smallholder. In phase two, female-headed households will be eligible for a 90 per cent grant.

The main elements of the package are:

- draught power (oxen or donkeys);
- implements (ploughs, row planters, cultivators, harrows);
- fencing (for arable land);
- water catchment tanks;
- scotch carts (introduced in 1991);
- threshing machine (being tested);
- grain silos (a new component for phase two).

Since 1988, there has been a policy of developing ALDEP demonstration farms with a farmer using a full complement of the package. These are planned to increase during phase two, to at least two in each of the 275 extension areas.

A strength of ALDEP was a deliberate strategy to combine research with practice. However, a study indicated that the attitudes of researchers and demonstrators to farmers inhibited the implementation of ALDEP – farmers were not regarded as resource people who could inform researchers and extension workers on the nature and extent of their problems. Farmers' problems were approached mechanically, with 'book solutions', even when these solutions were least relevant (BIDPA, 1996).

Reviews of ALDEP have noted the restrictive nature of some of the eligibility criteria – for instance, those without livestock were not eligible for fencing. This restriction was relaxed in 1989. However, in order to promote weeding, it has been obligatory since 1989 to take both a planter and a cultivator. The terminal report of the first phase records that farmers who received an ALDEP package were cultivating larger areas and producing more crops than those who did not receive a package. It is difficult to ascertain whether those receiving packages were already the keener farmers, and therefore it is hard to assess what the actual impact of the package has been. ALDEP is a very important programme, in that it has long term experience in trying to tackle difficult problems, using a combination of technologies, in a difficult environment. Further monitoring of impact is necessary.

Source: EDC,1997a

similar marketing conditions as those experienced by white farmers in the commercial areas). The future status of the company, including the possibility of it becoming a parastatal, is still under discussion. The operation of

MeatCo in the northern communal areas has resulted in a large increase in official sales (see Table 11.2).

Table 11.2: *Official Cattle Sales in Northern Communal Areas of Namibia*

Year	Cattle Sales
1983–91	4,000–5,000
1992	17,000
1995	29,690

Source: EDC, 1997d

CONCLUSION AND RECOMMENDATIONS

(1) Sustainable interventions to reduce market failure for smallholders need to be encouraged. Possible interventions include:
  • support to farmer organisations;
  • set-up support to improved marketing and input retailing;
  • investment in appropriate rural infrastructure;
  • support to contract culture;
  • appropriate market regulation which provides protection but not unnecessary restrictions for the smallholder;
  • improved financial services;
  • market information.
  Interventions need to be designed to avoid perpetuating the bias against sustainability and to include the poorest smallholders.

(2) Schemes which subsidise recurrent external inputs may further exacerbate the bias in favour of purchased inputs. They may well create a dependency on purchased inputs, which could result in additional hardship if the subsidies are later withdrawn.
  Whether they are effective in relieving poverty depends on:
  • their cost effectiveness in comparison to other forms of income transfer (for example, pensions);
  • the ability of the state to effectively target those households most in need of the inputs.
  These issues need to be looked at on a case-by-case basis.

(3) Credit services are in transition, with government-linked programmes trying to assert financial discipline and, as a consequence, often restricting their activities to richer farmers and profitable cash crops. NGO programmes have increased their emphasis on cost covering; the challenge is to operate sustainably, while still providing services to the poorest farmers. A number of innovative approaches are being tried with some success:
  • different models of working through groups and community organisations;

- working through NGO or local private sector intermediaries (eg traders)
- combining both savings and credit;
- linking between traditional practices and the formal financial sector;
- combining support to both cash and food crops at the farm level.

(4) There is a danger of lenders and borrowers sharing similar risk for rurally based savings and loan institutions – links to larger institutions or insurance are needed to spread the risk.

(5) There is scope for exploring ways of supporting credit supplied through rural retail traders as an alternative to, or in addition to, group-based schemes – competition perhaps may spread the risk of failure and provide a disincentive to exploitative interest rates.

(6) There is a need to develop appropriate insurance mechanisms which could reduce smallholder risk and encourage longer term investment. There is also scope for more learning from experience; some national and regional networks of credit institutions are including this objective.

(7) Subsidised investment in farm capacity, as used in Botswana, is likely to be more sustainable than subsidies to recurrent inputs. Subsidies could provide the incentive farmers need to take the long term view. However, it is important that the investments being subsidised are relevant to diverse smallholder needs, which means that the programme needs to be run sufficiently flexibly to adapt to individual needs. Farm capacity subsidies also need to be developed alongside other initiatives that support sustainability, including more appropriate extension advice, community natural-resource management capacity and interventions to improve the enabling environment, such as marketing.

(8) Support to local (on-farm and community) storage, processing and marketing may be imported for smallholders who are squeezed out of liberalised markets.

(9) Developing a number of alternative input and marketing opportunities can empower farmers to negotiate from a position of strength – for instance, over whether and how to participate in outgrower schemes.

(10) Affirmative action is needed to complement what is not being done by the market. For instance:
- making sure techniques aimed at long term sustainability are demonstrated and promoted, as well as the high external-input techniques demonstrated by suppliers;
- supporting small grain, minority crop and open-pollinated seed development and distribution where this is important to farmers, but not attractive to the commercial sector; appropriate link-ups between commercial companies, seed growers, NGOs and government can make such interventions commercially attractive and sustainable.

Chapter 12

Coping with Drought

INTRODUCTION

All the countries in the region are subject to drought and there are fears that these might get worse with global warming. In addition to causing enormous human suffering, drought undermines sustainability in a number of ways:

- Households lose – or are forced to sell – their productive assets (eg livestock are sold or die, while other means of production may be sold or exchanged for food).
- Drought directly damages the environment – trees die, branches are cut for fodder, trees may be cut for sale to buy food, grass is grazed to nothing, leaving the soil vulnerable on the return of the rains.
- Fear of drought discourages investment in productive and sustaining activities – such as buying fertilizer, buying livestock and planting trees.

There is a welcome shift in policy, in at least some drier countries in the region, from dealing with drought as an exception that requires an emergency response, to a recognition that drought is a regular feature and requires planning to cope with. Namibia is undergoing a planning process looking at drought management, which may bear interesting results (see Box 12.1).

Emergency response to drought in the region has cost large amounts of money and diverted considerable extension and other government time, often with unsustainable results. Some argue that if similar quantities of money and effort had been spent on prevention and coping strategies, there would have been considerably more progress towards sustainability.

However, it is sometimes difficult to get donor funds for this transformation to sustainability (see Chapter 13).

There is a need to differentiate between those areas for which drought is a regular occurrence and those wetter areas where droughts are infrequent (but can be just as devastating) since the strategy for managing drought can be very different. For instance, in Malawi, following the drought of 1991/92, aid agencies encouraged the adoption of a drought-resistant sorghum variety. This did well in the drier years of 1993/94 and 1994/95 but has failed due to fungus attack in the wetter years since then (see Chapter 5). Widespread distribution of varieties such as this is probably more relevant in areas with frequent droughts than in much of Malawi where drought is currently infrequent (Carr, pers comm).

MANAGING DROUGHT SUSTAINABLY

An essential part of sustainable farming is reducing risk and dealing with difficult circumstances when they arise. Diversity is a key risk-reduction strategy – having different fields, mixed cropping, a range of planting dates, and a range of on- and off-farm enterprises. Farmers often quite rationally prefer hardy to higher-yielding but vulnerable crops.

Some risk can be reduced by timely information. Price information in advance can help farmers make rational decisions on planting and selling. Recent advances in weather forecasting in Southern Africa mean it is increasingly possible to predict whether the season is likely to be wet or dry (although warning of an El Niño drought in 1997/98 seems to have been misplaced). This may, in the future, enable smallholder farmers to make adjustments to cropping paterns and planting dates (NRI, 1996). However, the extent to which farmers can rapidly switch to different crops (because of the availability of seed and planting material), as a result of a forecast that the current year is likely to be particularly dry or wet, is limited.

Coping Strategies

Even with effective risk-reduction strategies, disastrous years (or series of years) are likely to be encountered by smallholder farmers. These are most common and severe for the poorest and most vulnerable. Coping strategies are needed to survive these periods and these are an integral part of sustainable livelihoods – quite often they make use of common property resources. It is essential that actions such as tenure changes to common land are analysed for their effect on coping strategies to avoid undermining a vital part of the livelihoods of the most marginalised part of the population. Women, in particular, need to be consulted. Coping strategies are location specific but generally include many of the following:

- reduction in expenditure (school fees, meat etc), reduced consumption;
- using up stored food reserves;
- using up cash reserves;
- increased use of traditional exchange mechanisms – work paid for with food and informal loans, etc;
- greater use of family networks and greater reliance on remittances;
- selling of livestock, trees and other possessions;
- greater use of famine foods and food from the wild;
- search for alternative employment;
- eating next year's seed;
- migration.

Household asset levels before a crisis are important in enabling them to survive the crisis. Therefore, poorer households are more vulnerable and economic conditions resulting in low asset levels increase general vulnerability. Policies which improve the overall livelihood level of the rural population, other things being equal, will reduce the need for drought relief. Agricultural policies have tended to promote increased production at the expense of drought resilience; affirmative action is needed to research and encourage strategies that maintain diversity and risk management (Chapter 4).

In most countries in the region, preliberalisation drought management at a national level consisted of national food stockpiles. With the recognition of the high costs of managing such stockpiles, and trends towards integrating within regional and world markets, there are moves towards scaling down stockpiles, developing regional stockpiles, and relying on being able to purchase food on the regional or world market in times of drought. While there is financial logic in this approach, there are also risks:

- A region-wide drought would draw down stock across SADC.
- Transport bottlenecks, particularly for landlocked countries, can delay imports.
- Lack of liquidity makes timely purchases on the world market difficult for some countries.
- There is a lack of guaranteed availability of grain on the world market.

This strategy was almost disastrous in 1991/92, when food deficits throughout the region caused acute import bottlenecks. The rehabilitation of the transport corridors through Mozambique may mean that this risk is reduced, but nevertheless many countries still try to keep a strategic reserve.

Drought relief, although sometimes essential, tends to externalise the cost of drought onto the government and donors and so, in turn, does not encourage risk-minimising behaviour by farmers. In some countries, drought relief has become an almost regular feature of the 1990s. Although this partly reflects the severe droughts experienced in the 1990s, it also reflects the general critical condition of many rural households and the political agenda of some politicians and donors. The reactive nature of such expenditure does not encourage the long term strategy required for encouraging sustainability.

Arguments against regular intervention in drought do not override the need for effective safety nets to help the poorest; the challenge is to ensure that assistance is better targeted to those with greatest need, and to design schemes which encourage, rather than discourage, sustainability.

Drought relief and recovery policies have not always been orientated towards encouraging sustainability:

- Drought recovery seed and fertilizer packs have tended to emphasise bought inputs, and the seed has often been maize, even in traditional small-grain growing areas (see Box 11.5). Cash may be more appropriate than either food or Agpacks, empowering households to decide their own priorities and how to meet them.
- Drought relief has not often been developed with the participation of local communities, thereby undermining self-reliance and missing opportunities for locally sustainable solutions. There are exceptions – for example, in the 1991/92 drought in southern Zimbabwe, NGOs and government provided bought feed for a limited number of breeding animals that had been selected by each village, thereby protecting the asset base for recovery and involving the community in deciding how to manage the programme (Mombeshora et al, 1995).
- Aid agencies and governments are often poor at informing communities about plans for relief deliveries, yet information like this is vital for households who need to take momentous decisions – for instance, about whether to sell their last chicken or cow to buy food.
- Drought relief programmes, such as food-for-work, can be an opportunity for increasing organisational capacity in the community. Unfortunately, too often they create dependency and undermine the capacity that already exists. This is often because such programmes are organised too late and are driven from above – by politicians and large donor organisations, rather than building on local initiatives and long-term relationships between communities and more locally based organisations.

Given the devastating impact of recent droughts on livestock ownership by poorer farmers, there is a need to explore ways of encouraging rapid destocking at the beginning of a drought – before animals lose condition and value, and die – and facilitating restocking after a drought. This requires:

- a transparent and secure system, with appropriate community participation which farmers understand and have confidence in;
- favourable livestock purchase prices being maintained on the onset of drought;

Box 12.1 Namibian Drought Policy

A general perception has developed over the past few decades that droughts, warranting government-funded relief, are frequent events. Farmers have been receiving substantial drought relief assistance from the government since the 1950s. In the seven years since independence, livestock farmers (often amongst the wealthier members of society) have received an estimated N$315 million (US$72 million) in livestock-related assistance alone (National Drought Task Force, 1997). Furthermore, the readiness with which government has been willing to provide assistance is increasingly recognised to have discouraged farmers from adopting management practices that are adapted to Namibia's natural aridity. On the contrary, a number of interventions, such as the subsidy on fodder, have led to unsustainable practices by encouraging farmers to keep stock when they should be marketed. This, in turn, inhibits the recovery of grazing. These interventions not only promote unsustainable farming practices but are highly inequitable, with the bulk of assistance being captured by wealthy livestock owners and commercial maize farmers, rather than the poor.

A new policy is currently being discussed which aims to shift responsibility for managing drought risk from the government to the farmer (MAWRD, 1997). It is now argued that previous perceptions of the degree of drought which warranted relief measures were misconceived. Drought relief should only apply when a disaster drought, defined in terms of its extremity and rarity in relation to normal arid conditions, occurs. The policy seeks rather to emphasise long term measures that support the management of risk by farmers. Short term assistance to support crop and livestock farmers would be limited to disaster droughts when conditions are so severe or protracted that they are beyond what a farmer would be expected to deal with, in terms of normal risk management.

This calls firstly for an objective definition of drought, which cannot be subjected to the dictates of political expediency. Relief assistance to commercial farmers should in future be financed mainly by farmers themselves and industry levies. Policy reform will promote sustainable farming practices by supporting mechanisms which will stabilise their incomes under conditions of extreme climatic variability. Reforms are needed in the tax regime in which farmers operate, and incentives to promote investment in diversified farming and non-farming operations are required. For equity reasons, communal farmers will continue to receive government-financed drought relief. Support for communal-area livestock farmers in disaster droughts should take the form of livestock marketing incentives and, in extreme cases, the preservation of core breeding stock through financial support to the leasing of and transport of livestock to emergency grazing.

It is also proposed that long term programmes must be instituted to ensure that all communal-area crop farmers have access to – and can afford the inputs for – production in the year after drought, flood or pest damage. Free seed handouts could be abolished on condition that seeds are readily available and affordable. The agricultural input voucher scheme, as discussed in Box 11.11, could be pursued and designed so that it can be upgraded after drought years to compensate poorer (agriculture dependent) farmers for the loss of vital inputs (eg seeds) and other assets (eg draught animals and implements) suffered as a result of droughts.

Source: EDC, 1997d

- increased slaughter and cold storage facilities;
- purchase of old or poor condition livestock for famine relief;
- the ability to target poor households for preferential destocking support;
- possibly a community-based system for maintaining a core breeding herd;
- alternative savings opportunities for farmers;
- a system of restocking (such as rotating loan schemes for livestock) – although, if the drought and livestock mortality is widespread it may be difficult to procure livestock for restocking.

Such schemes would be expensive and they would tend to externalise the cost of drought away from the farmer (possibly encouraging overstocking). Therefore, caution and targeting are needed. However, given the devastating impact of drought on smallholder livestock owners, schemes like these warrant greater attention.

CONCLUSION AND RECOMMENDATIONS

Drought plays a major role in undermining agricultural sustainability in Southern Africa:

- There is a need to move away from emergency responses to invest in sustainable forms of agriculture, which are able to produce in drought-prone environments.
- Post-drought 'recovery inputs' need to be appropriate for sustainable farming in drought-prone areas – with less emphasis on maize seed and fertilizer.
- When drought relief is essential, there must be greater community involvement in the process, and more information, in order to empower farmers to cope with drought in a way that does least to undermine their long term sustainability.
- There is a need for more action research on appropriate destocking and restocking of livestock within tracking management systems.
- The importance of wild produce in times of drought needs to be recognised and common property entitlements conserved.

Part V

The Way Forward

Chapter 13

Short Term Financial Balance or Long Term Sustainability and Poverty Eradication?

Donor pressure has been instrumental in persuading many governments to rapidly phase out agricultural subsidies and other controls, often in the name of 'sustainability'. However, what is being introduced in these cases is the financial sustainability of agricultural services, rather than the sustainability of smallholder agriculture. One does not necessarily lead to the other – in some cases too rapid a move to programme financial sustainability may undermine agricultural sustainability, or cause immense amounts of human suffering.

It is often argued that subsidies:

* are inherently unsustainable because of the costs involved;
* create distortions, thereby reducing efficiency and raising costs.

Certainly, the comprehensive government intervention in some countries spawned inefficiencies, unsustainable parastatals and corruption – and often did not help the poorest farmers (see Chapter 3). However, critics point out that agricultural markets are controlled and subsidised in many rich countries, and have been so for considerable lengths of time. Southern African farmers are expected to adapt to an unsubsidised market-led environment in an unfairly short time period, when many northern farmers continue to receive subsidies and other support.

It can be argued that Southern African agriculture is in a state of transition, with rapid population growth, rapid urbanisation, rapid introduction of various technologies and rapid changes in the overall political and economic environment. African farmers are expected to achieve a transition to sustainable intensification, yet the goal posts are continually being moved, with new policies, prices and programmes coming and going, undermining the stability and security which are part of those ingredients needed to persuade farmers to invest in long term sustainability.

It is important to differentiate between:

- the physical sustainability of farming (maintaining soils and biodiversity, etc);
- the basic needs of the rural population;
- the financial sustainability of agricultural services, programmes and policies.

The physical sustainability of farming has to be maintained now to avoid even greater problems in the future. Human needs must be met to avoid greater impoverishment than is already present in many rural areas, and to create the conditions which encourage sustainable rates of population growth. The financial sustainability of programmes and policies is more negotiable. Clearly, in the long term they need to be sustainable; but if a transition is taking place, then the wisest policy may be to finance this transition, even if it means delaying reaching fiscal balance. There are precedents – neither Roosevelt's New Deal in the US, nor the Marshall Plan in post-war Europe were sustainable in themselves, but they financed a transition that was considered necessary.

A problem is that timescales given to African farmers and African governments for transition are often unrealistically short. Indications are that, if the policy environment is appropriate, rates of population growth will come down and begin to stabilise. This will make the attainment of sustainable agriculture much more realistic and will, in turn, enable agricultural programmes and policies to be financially sustainable. However, this process is likely to take decades rather than years. The danger is when the rush to financial sustainability in the short term destroys either physical sustainability or causes increased poverty – exacerbating suffering and undermining reductions in birth rates. This is exactly what is in danger of happening in much of Southern Africa.

Warning against overrigid and overrapid insistence on financial sustainability does not mean giving a blank cheque for agricultural subsidies. It does, however, mean that some long term programmes and policies will be required that:

- support farmers to develop long term, sustainable techniques for increasing the production from the land;
- relieve poverty during a lengthy transition phase, without creating a bias against sustainability and self-reliance.

Such programmes and policies will need long term government and donor support. Political will is necessary, alongside evidence that the proposed actions will really encourage a transition to sustainability rather than dependency.

Chapter 14

Conclusions and Recommendations

INTRODUCTION

Smallholder agriculture remains the major source of livelihood for most of the rural poor, yet it is failing to provide a route out of poverty for the majority. Although opinions differ about the severity of environmental problems, current agricultural practices in much of the region are proving unsustainable, particularly in relation to maintaining soil fertility and managing common property resources.

TECHNOLOGICAL BIAS

Most government research programmes in the region still concentrate on short term, single-crop yield maximisation and increased use of external inputs. Often recommendations are too expensive to be followed by poorer smallholders. The reasons for this bias are quite logical:

- Research and extension on low external-input technologies tend to be long term and more complex, and therefore tend to be neglected.
- Most of the agricultural establishment has been trained within the yield maximising, high external-input ethos.
- An increasing proportion of research and demonstration is being performed by input suppliers, who naturally emphasise the use of bought inputs.
- Farmer organisations, who may advocate more research, often consist mainly of those smallholders with sufficient resources to use more inputs.

- NGO research is expanding and may be more orientated towards sustainability, but it is still very limited and often lacks a systematic and long term perspective.

This bias has led to an underdevelopment and underpromotion of potentially sustainable alternatives to current practices. Government policy in much of the region is moving in favour of commercialisation of smallholder farming; this is likely to reinforce past biases by concentrating on more purchased inputs and cash crops – which runs the risk of neglecting the needs of the poorer smallholders and giving insufficient attention to the issue of sustainability.

Publicly funded research programmes need affirmative action in favour of sustainability to redress the current bias. Paradoxically, due to the very low use of external inputs (such as inorganic fertilizer) in the region, there is generally scope for increased use of external inputs as well as a much greater emphasis on low external-input technologies. It is not an either/or situation; both are needed to achieve sustainable intensification.

To achieve affirmative action on equity and sustainability within public-sector agricultural research there need to be:

- clear priorities in favour of smallholders and sustainability;
- a long term perspective to make possible experiments into maintaining and enhancing yield over the longer term (this implies a degree of funding security);
- active creation of a learning environment involving farmers, extension and research, with research and extension stimulating, enabling and publicising research done by farmers;
- options for farmers to choose from rather than blueprints;
- appropriate reward systems, rewarding researchers for adoption and sustainability of technologies developed rather than for yields obtained on station and papers published;
- research planned in accordance with the priorities of smallholder farmers, including women, and within an understanding of the overall farming system;
- adaptive locally based research, responsive to diverse environments;
- research, not only looking at how to increase yield, but also how to reduce costs, reduce labour, reduce risk and reduce environmental damage;
- a multi-disciplinary approach making use of multidisciplinary teams.

There is a need to address both the potential benefits and the potential risks from biotechnology with affirmative action to:

- Ensure the interests and opinions of Southern African smallholders, consumers and companies are listened to when regulating biotechnology activities.

- If appropriate within these regulations, ensure (probably with public-sector funding) that the priorities of Southern African smallholders are included in future biotechnology research.

There are a number of resource-conserving techniques that seem able to improve smallholders' yields at minimal cost; most of these tend to be quite location specific. We should not expect a magic bullet – instead, a number of techniques will, in most cases, be required to achieve sustainability. Barriers to adopting resource-conserving technologies need to be researched and overcome.

SUSTAINING LOCAL INSTITUTIONS

Local institutions can support sustainable agriculture by:

- mediating between households and community-held entitlements;
- influencing norms of behaviour (eg on cultivation practices and cutting trees);
- undertaking large activities, such as water development, feeder road maintenance and market development;
- lowering transaction costs for services;
- managing common property resources;
- lobbying for policies and services appropriate to their needs.

Actions that increase the capacity and confidence of communities and farmer groups to do these things can provide the preconditions for sustainable agricultural development. However, in enthusiastically trying to develop community capacity, it should be remembered that community organisations are not necessarily homogeneous, democratic or free of gender bias.

There are many different options for community-based agricultural interventions, such as:

- whether to pay compensation or salaries to community-based intermediaries;
- what role traditional leaders should play, and what type of relationships to have with government institutions;
- whether to work with single issue or generic groups;
- optimum size, tiers or structures for groups;
- whether to work through existing structures or to create new groupings.

The final choice needs to be made in participation with the community, so blueprints are inappropriate. Decentralisation of local government provides new opportunities to develop alliances and to support locally appropriate and accountable services. Although participation is increasingly recognised as a vital component of development, there are many different types and levels, and there are also some costs – both to community members in time taken up and to the agency in facilitation. Therefore, agencies need to understand the local situation and to consider various options, including working through existing informal organisations or encouraging local entrepreneurs to provide services, before engaging in group formation.

Participation is both an approach and a process; the right attitudes are needed as well as the skills. Participatory development is neither a short term process nor one where the outcome can be defined in advance; this means that facilitators and donors require:

- long term commitment to an area and a community;
- more flexibility than is often encouraged by the use of logical frameworks, with development plans evolving to meet the developing needs of the community.

Those bringing in participatory techniques must build up their experience and be aware that participation can not only bring great benefits, but can be misused for narrow factional ends. Field workers need to be trained to combine technical agricultural knowledge with the experience of working with local communities.

Farmer associations and farmer unions form part of this picture; they can act as a link to services, be service providers in their own right, or can lobby for improved services and policies. It is, however, unreasonable to expect a smallholder union, representing a diverse membership, to necessarily give priority to the issue of long term sustainability and the concerns of the poorest smallholders. Affirmative action and alliances with other organisations are necessary to achieve this.

In much of Southern Africa the sustainability of, and access by the poor to, vital common-property resources is under threat from unregulated use and privatisation by the better off (eg fencing of grazing land). A range of innovative community-based natural resource management (CBNRM) programmes, mainly for wildlife management, are being developed. Although for many communities the advantages are not as great as some proponents suggest, these still represent a major step forward in combining sustainable resource management with more immediate human needs.

There is a need to scale up and broaden out from wildlife based programmes to a more holistic approach, including grazing, timber, veld products and fish. This will involve:

- *Legal and policy changes* – which are sometimes needed in order to include a wider range of natural resources in CBNRM. Land law and CBNRM law need to be compatible, giving communities tenure security over common land and the natural resources associated with the land. Communities need to be able to punish offenders and exclude outsiders from resource expropriation – traditional rules must be backed by law and state enforcement support.

- *Awareness raising* – a more generalised transformation of the policy and attitude in support of CBNRM is required. This necessitates more awareness among local politicians, and also among many government officials from departments outside of wildlife management who are concerned with resources such as water, trees, soils and grazing.
- *Community capacity building* – this includes leadership training and efforts to empower women.
- *Giving value to environmental resources* – consideration should be given to levies on environmental resources transported out of rural areas, such as firewood, charcoal and thatching grass – to be charged by and paid to local communities. This right can be made dependent on implementing a sustainable management plan, and could provide an incentive for sustainable resource management..
- *Greater recognition of conflicting interests* within communities and a willingness to study and research these – typically, these may be between rich and poor households or between women and men, but can also have other dimensions, such as age or ethnicity.
- A *learning approach* is needed in which experience gained in early schemes can be built upon. In particular, new programmes being developed under the environmental banner, (such as landuse planning initiatives) need to be aware of the experience already gained in past agricultural and integrated rural development projects.

There is scope for building more innovative partnerships with traditional and spiritual leaders over environmental matters; however, this will mean real partnership, rather than using these leaders as a cheap way of implementing an externally set agenda.

There is considerable controversy in the region on the degree of overgrazing, the causes and potential solutions. Some livestock departments in the region are sceptical about the sustainability of communal grazing and have not yet embraced some of the newer thinking on tracking management. However, sustainably managed communal schemes are also rare. This is a critical area which needs more research and a more open and critical exchange of experience.

CREATING AN ENABLING EXTERNAL ENVIRONMENT

Probably the most important yet elusive factor in encouraging sustainability is confidence – confidence in the future to make it worth investing in. In recent years the sheer pace of change in the economic and social environment within which agriculture operates has undermined any long term perspective; confidence needs to be rebuilt.

Land tenure security is an essential incentive for farmers to invest in long term sustainability. Often too little is known about current land

tenure, how this is changing, and its interaction with sustainable management. Traditional forms of tenure are not static but are evolving in
response to new realities. The direction of this evolution can be influenced
and communities need to be supported to develop systems which encourage sustainability and access to land by women. Similar advocacy for
sustainability, women and the resource poor is needed if systems are
changing from traditional to formal (eg titling) arrangements and if a land
market develops. Common property rights are under threat in many areas
and community authority needs to be reinforced, with recognition and
legal back-up for exclusive group rights.

The lessons from the land reform programmes in the region are:

- Settlers need tenure security.
- A more participatory planning process is required, taking into
  account diversity of objectives and circumstances.
- More attention needs to be paid to institutional development,
  especially for the management of common property resources; overlarge groups should be avoided.

Government agricultural extension has tended to suffer from the same
biases as research, favouring slightly better-off farmers and simple yield
maximisation rather than a range of choices more relevant to resource-
poor farmers. The very large gap in all countries between extension
recommendation and small farmer practice is an indication of poor
research and extension performance.

There are initiatives across the region aimed at improving the management of ministries of agriculture by making decisions more transparent
and demand driven (eg Zimbabwe's Agricultural Services Management
Project and Zambia's Agricultural Services Investment Project). While
projects like these may be able to provide the preconditions for a transformation in favour of smallholder sustainability, they will not, in themselves,
do so without specific affirmative action. Some of the changes needed are:

- Affirmative action in favour of sustainability – demonstrating and
  promoting a combination of resource-conserving practices.
- Creation of a learning environment at the smallholder–extension
  interface – this implies the promotion of choices, the encouragement
  of farmer experimentation and the facilitation of farmer-to-farmer and
  farmer-to-research exchanges.
- Decentralised organisation – giving local managers more flexibility to
  organise extension, according to local conditions. The trade-off is that
  local managers would need to be accountable for their local impact.
  Increased accountability to local farmers, impact indicators sensitive
  to sustainability and poverty criteria are also needed.
- Collaboration with NGOs and CBOs: much collaboration to date has
  been ad hoc and short term – there needs to be a realisation that the
  involvement of local NGOs and farmer associations in extension can

be effective and long term, and thus worthy of strategic development. There is scope for exploring the possibilities of more contracting-out of extension tasks to local NGOs or farmer associations. However, NGOs cover a wide variety of competencies and more independent and systematised evaluation is needed. Some that appear successful at a small scale are too expensive to be scaled up or replicated.

- Collaboration with the commercial sector – this may range from providing extension to outgrower schemes and collaborating in demonstrations on the use of specific inputs, to the promotion of particular commercial products. There are both opportunities and dangers in this collaboration; guidelines need to be drawn up to prevent bias and public-funded extension must be orientated to fill gaps left by the commercial sector.
- Divestment of 'non-extension' tasks – divestment of some of these tasks can provide opportunities for redefining core extension roles and creating an optimum allocation of responsibilities between a range of government, commercial and non-profit organisations.
- A more fundamental shift of emphasis is needed within formal training institutions – this should be coupled with improved in-service training in order to produce people with the right skills and attitudes to effectively support sustainable smallholder farming.

Rural financial services – savings, credit and insurance – can encourage sustainability by:

- facilitating rational farm management by acting as a buffer between cash needs and optimum income-generating strategy;
- providing resources for seasonal or longer term investment;
- reducing risk (through diversified savings or insurance), thereby encouraging longer term planning and investment.

In the past, financial services have been dominated by inequitable and unsustainable state-run credit schemes, reinforcing existing bias by subsidising the use of external inputs. Credit services are in transition, with government-linked programmes trying to assert financial discipline and, as a consequence, often restricting their activities to richer farmers and profitable cash crops. NGO programmes too have increased their emphasis on cost covering; the challenge is to operate sustainably, while still providing services to the poorest farmers.

A number of innovative approaches are being tried with some success:

- different models of working through groups and community organisations;
- working through NGO or local private-sector intermediaries (eg traders);
- combining both savings and credit;
- linking between traditional practices and the formal financial sector;

- combining credit to both cash and food crops at the farm level.

There is a danger of covariant risk for rurally based saving and loan institutions; links to larger institutions or insurance are needed to spread the risk.

There is scope for exploring ways of encouraging credit supplied through rural retail traders as an alternative to, or in addition to, group-based schemes – competition can spread the risk of failure and provide a disincentive to exploitative interest rates. There is also a need to develop appropriate insurance mechanisms which could reduce smallholder risk and encourage longer term investment. There is scope for more learning from experience; some national and regional networks of credit institutions are developing this objective.

Drought plays a major role in undermining agricultural sustainability in Southern Africa:

- There is a need to move away from emergency responses, to invest in sustainable forms of agriculture which are able to produce in drought-prone environments.
- Post-drought recovery inputs must be appropriate for sustainable farming in drought-prone areas – with less emphasis on maize seed and fertilizer.
- When drought relief is essential, there needs to be greater community involvement in the process and more information, in order to empower farmers to cope with drought in a way which does least to undermine their long term sustainability.
- There is a need for more action research on appropriate destocking and restocking of livestock, within tracking management systems.
- The importance of wild produce in times of drought must be recognised and common property entitlements conserved.

Sustainable interventions to reduce market failure for smallholders need to be encouraged, particularly those that enable farmers to use fertilizer profitably and sustainably. Possible interventions include:

- support to farmer organisations;
- set-up support for improved marketing and input retailing;
- investment in appropriate rural infrastructure;
- support to contract culture;
- appropriate market regulation, which provides protection but not unnecessary restrictions for the smallholder;
- market information;
- improved financial services.

Interventions need to be designed to avoid perpetuating the bias against sustainability and to include the poorest smallholders. Development of a number of alternative input and marketing opportunities can empower farmers to negotiate from a position of strength – for instance, over whether and how to participate in outgrower schemes.

Schemes which subsidise recurrent external inputs may further exacerbate the bias in favour of purchased inputs, noted throughout this book. They may well create a dependency on purchased inputs, which could result in additional hardship if the subsidies are later withdrawn. Whether they are effective in relieving poverty depends on:

- their cost effectiveness in comparison to other forms of income transfer (for example, pensions);
- the ability of the state to effectively target those households most in need of the inputs.

These issues need to be looked at on a case-by-case basis.

Subsidised investment in farm capacity, as used in Botswana, is likely to be more sustainable than subsidies to recurrent inputs. Subsidies could provide the incentive farmers need to take the long term view. However, it is important that the investments being subsidised are relevant to diverse smallholder needs, which means that the programme must be run sufficiently flexibly to adapt to individual needs. Farm capacity subsidies also need to be developed alongside other initiatives supporting sustainability, including more appropriate extension advice, community natural-resource management and interventions to improve the enabling environment, such as marketing.

REDEFINING ROLES IN SUSTAINABLE AGRICULTURAL SERVICE PROVISION

Agricultural services can be provided by a mix of government, commercial, NGO and community organisations. The appropriate mix needs to be developed locally and pragmatically, depending on:

- the relative strengths of local organisations (which are likely to change over time);
- the different comparative advantages organisations have for different tasks.

A balance of competition and collaboration is needed between the different service organisations.

Although government is unlikely to return to the dominant role aspired to in the 1970s and 1980s, it needs to:

- Create the enabling environment for other service providers.
- Redress imbalance in existing service provision by affirmative action in favour of sustainability and the poorest smallholders.
- Regulate potentially dangerous practices.

This requires developing resources and new skills within government, ideally at a decentralised level. Programmes such as the Rural District Capacity Building Programme in Zimbabwe are trying to do this.

Sustainability requires a long term perspective which government, like the other stakeholders, finds difficult to achieve. Government can help to create the stable conditions that encourage other stakeholders to take a long term view. Donors can also support initiatives within government in favour of the longer term and sustainability; this means donors have to take a longer term view as well.

A major task is to persuade policy-makers within Southern African governments, and in powerful donor institutions such as the World Bank, that current policies are neither ending poverty nor protecting the environment, and that there is a viable alternative. This alternative requires progress on three fronts:

• more development and promotion of a range of resource-conserving and productivity-enhancing technologies, which farmers can choose from and experiment with;
• more organisational development at the community level, and policies giving power to these organisations to create a local environment favouring agricultural sustainability;
• wider policies and programmes that provide security and incentives for farmers to invest in long term sustainable and productive agriculture.

Those of us promoting this alternative need to build up evidence to show that it can be successful. This requires improved monitoring and evaluation of existing experience, including more transparent sharing and analysis of impact at all levels.

Appendix 1

Some Useful Addresses

This is not a complete list but gives addresses of some useful organisations and of some of the projects mentioned in the text.

BOTSWANA

Cooperation for Research, Development and Education (CORDE)
PO Box 1895, Gaborone, Tel: (267) 323865

Forum on Sustainable Agriculture (FONSAG)
P Bag BO 136, Bontleng, Gaborone, Tel:/Fax: (267) 301971

Ministry of Agriculture
P Bag 003, Gaborone. Fax: (267) 356027

National Conservation Strategy (Coordinating) Agency
P Bag 0068, Gaborone, Botswana, Tel: (267) 302050/302056.
Fax: (267) 302051

Rural Industries Promotions Company (RIP)
PO Box 2088, Gaborone, Tel: 314431/2. Fax: 300316

Thusano Lefatsheng
PO Box 00251, Gaborone, Tel: (267) 3722273

Veld Products Research
PO Box 2020, Gabarone, Tel:/Fax: (267) 347047

Women's NGO Coalition
P Bag 342, Gaborone, Tel:/Fax: (267) 30955

MALAWI

Coordination Unit for the Rehabilitation of the Environment
PO Box 1429, Blantyre, Tel: 621 451; Fax: 621 468;
E-mail: cure.malawi@unima.wn.apc.org
Contact point for other NGOs working on natural resource management;
has a resource centre and produces a newsletter

Malawi Agroforestry Extension Programme
PO Box 30291, Lilongwe 3, Tel: 741 988, 741 986; Fax: 744 064
Produced manual on agroforestry practices in Malawi

Ministry of Agriculture and Livestock Development
PO Box 30134, Lilongwe 3

Ministry of Forestry and Natural Resources
PO Box 30048, Lilongwe 3

Ministry of Research and Environmental Affairs
PO Box 30745, Lilongwe 3

Poverty Alleviation Programme Pilot Programme Agroforestry
PO Box 1481, Lilongwe

Rockefeller Institute
PO Box 30721, Lilongwe 3

SADC–ICRAF Agroforestry Project
Box 31300, Lilongwe 3
Regional office with projects in several countries.

NAMIBIA

Desert Research Foundation of Namibia
PO Box 20232, Windhoek, Tel: +61–229855; Fax: +61–230172

Directorate of Environmental Affairs, Ministry of Environment and
Tourism
Private Bag 13306, Windhoek, Tel: +61–249015; Fax: +61–240339

Integrated Rural Development and Nature Conservation
Katima Mulilo, Tel: +677–2108; Fax: +677–3453
c/o PO Box 1715, Swakopmund

Likwama Farmers Cooperative Union
PO Box 179, Katima Mulilo, Tel: and Fax: +677 3561

Ministry of Agriculture, Water and Rural Development
Private Bag 13184, Windhoek, Tel: +61–2022146

Namibia National Farmers Union
PO Box 50364, Windhoek, Tel: +61–271117

Sustainable Animal and Range Development Programme
MAWRD, Private Bag 13184, Windhoek, Tel: +612022146

SOUTH AFRICA

Agricultural Research Council
Box 8783, Pretoria 0001, Tel: 012 3429968; Fax: 012 434912

Agriculture and Rural Development Research Institute (ADRI)
University of Fort Hare, Post Bag X1314, Alice, 5700, Tel: 0406 531154;
Fax: 0406 531730

Environment and Development Agency (EDA)
Box 322, Newtown, 2113, Tel: 011 8341905; Fax: 011 836 0188

Farmer Support Group
Post Bag X01, Scottsville 3209, Tel: 0331 2606281; Fax: 0331 260 6281

Institute of Natural Resources (INR)
University of Natal, Post Bag X01, Scottsville 3209, Tel: 0331 460796; Fax:
0331 460895

Land and Agricultural Policy Centre (LAPC)
Box 243, Wits 2050, Tel: 012 4037272; Fax: 012 3396423

ZAMBIA

Carc Zambia
Box 60256, Livingstone, Tel: Livingstone 324259

Clusa
c/o USAID, Box 32481, Lusaka, Zambia, Tel: 01–254303/6

Conservation Farming Unit
c/o Zambia National Farmers Union, PO Box 30395, Tel: 01–262315

DAPP Zambia
Box 70505, Ndola, Tel: 02–615391/615491

Farming Systems Research Team (formerly ARPT)
Ministry of Agriculture, Food and Fisheries, Mt Makulu Central Research
Station, Post Bag 7, Chilanga, Tel: 01–278008/255263

HODI
Box 36548, Lusaka, Tel: 01–290455

Institute for Cultural Affairs of Zambia
Box 31454, Lusaka

Kaloko Trust
PO Box 71737, Ndola

Luano Valley Development Committee
c/o Chingombe Catholic Mission, Private Bag 8012XK, Kabwe

Namwala Cattle Project
Dept of Animal Production and Health, Box 3, Namwala

Ndola Rural Technician and Smallholder Farmers' Association
District Forestry Office, Box 18, Masaiti, Copperbelt Province

ZIMBABWE

Africa Centre for Holistic Resource Management
PO Box MO 266, Mt Pleasant, Harare, Tel/Fax: 776942;
E-mail: <root@achrm2.icon.co.zw

Agritex
PO Box CY 639, Causeway, Harare, Tel: 794601; Fax: 730525

CARE Zimbabwe (AGENT Programme)
PO Box HG 937, Highlands, Harare, Tel: 727986/7/8; Fax: 727989

Department of Research and Specialist Services
PO Box 8108, Causeway, Harare, Tel: 704531; Fax: 728317

ENDA Zimbabwe
PO Box 3492, Harare, Tel: 301156/301162

Fambidzani Permaculture Centre
PO Box CY 301, Causeway, Harare, Tel: 307557/726911; Fax: 726911;
E-mail: fambidzani@mango.zw

ICRAF
c/o DRandSS, Box CY 594, Causeway, Harare, Tel: 704531; Fax: 728340;
E-mail: icraf-zimbabwe@cgnet.com

IUCN (Regional Office for Southern Africa)
PO Box 745, Harare, Tel: 737951/2,/3; Fax: 791231

National Association of Cooperative Savings and Credit Unions of
Zimbabwe
Box CY 176, Causeway, Harare, Tel: 793815; Fax: 739610

Self-Help Development Foundation of Zimbabwe
Box 4576, Harare, Tel/Fax: 572933
Silveira House
E-mail: Silveira@healthnet.zw

Zimbabwe Farmers' Union
Box 3755, Harare, Tel: 772859/60; 772863/4; Fax: 750456

Appendix 2

Sources of Information on Sustainable Agriculture

There are many books and different sources of information. Those given below are the ones the author has found most useful as a background to the issues dealt with in this book.

BOOKS

Whose Reality Counts? Putting the First Last. Robert Chambers, 1997. IT Publications.* A challenging book celebrating the capacity of local communities to analyse and plan, and warning of the arrogance and mistakes made by professionals and the powerful.

Living with Uncertainty – New Directions in Pastoral Development in Africa. Edited by Ian Scoones, 1995. IT Publications.* Examines the management and policy implications of the new thinking on environmental variability over time and space and the role of pastoral societies in managing these.

Regenerating Agriculture – Policies and Practice for Sustainability and Self-Reliance. Jules Pretty, 1995. Earthscan Publications. Examines how many farmers are still locked into modernist approaches to agriculture, which are dependent on high levels of external inputs, and examines alternative sustainable agricultural practices.

---

* These and many other books (including many not published by IT) are available by post from IT Publications, 103–105 Southampton Row, London WC1B 4HH, UK, Tel: 44(0) 171 436 9761; Fax: 44(0)171 436 2013. E-mail: itpubs@gn.apc.org.

Beyond Farmer First – Rural People's Knowledge, Agricultural Research and Extension Practice. Edited by Ian Scoones and John Thompson, 1994. IT Publications.* Looks at some of the experiences and practical implications of putting farmers first in agricultural development programmes.

The Southern Africa Environment – Profiles of the SADC Countries. Edited by Sam Moyo, Phil O'Keefe and Michael Sill, 1993. Earthscan Publications. Provides information on physical and human geography, environmental problems, resource base and institutional structures for environmental management.

Hazards and Opportunities – Farming Livelihoods in Dryland Africa: Lessons from Zimbabwe. Ian Scoones et al, 1996. Zedbooks Ltd. 7 Cynthia St, London N1 9JF. Detailed analysis of some of the processes taking place in a dryland area of Zimbabwe makes this book relevant to a much wider audience.

Participatory Learning and Action – A Trainer's Guide. Pretty, Guijt, Thompson, Scoones, 1995. International Institute for Environment and Development (IIED), 3 Endsleigh St, London WC1H 0DD, Tel: 44(0)171 388 2117; Fax: 44(0)171 388 2826; E-mail: iiedagri@gn.apc.org Practical guide to training others in the use of participatory methods.

A Field Manual for Agroforestry Practices in Malawi. Bunderson et al, 1995. Malawi Agroforestry Extension Project, PO Box 30291, Lilongwe 3, Malawi. Although written for Malawi, this manual has wider relevance, particularly to those parts of Southern Africa with higher rainfall.

Agricultural Extension. Van de Ban and Hawkins, 1996. Blackwell Science Ltd. A second edition of a basic text on extension, which includes some of the newer thinking on sustainability and plurality of service providers.

Sustaining the Soil – Indigenous Soil and Water Conservation in Africa. Edited by Chris Reij, Ian Scoones, and Camilla Toolmin, 1996. Earthscan Publications. Documents farmer practices and adaptations developed over generations in response to changing circumstances.

Farming for the Future: An Introduction to Low External-Input and Sustainable Agriculture. Coen Reijntjes, Bertus Haverkort, Ann Waters-Bayer, 1992. Macmillan Publishers. Explores how development workers can assist small-scale farmers in making the best use of low-cost local resources to solve their agricultural problems.

Training for Transformation – A handbook for Community Workers. Vols I,II,III. Anne Hope and Sally Timmel, 1984. Mambo Press, Box 779, Gweru, Zimbabwe. Handbooks to accompany Training for Transformation courses used in several countries in the region.

Microfinance and Poverty Reduction. Susan Johnston and Ben Rogaly, 1997. Oxfam Publications, 274 Banbury Rd, Oxford, OX2 7DZ, UK. A useful introduction into how savings, credit and other financial services can be used to improve the livelihoods of poor people.

NETWORKS AND PERIODICALS

Arid Lands Network (ALIN): CP 3, Dakar-Fann, Senegal. Produces newsletter three times a year exchanging information between those working in the arid parts of Africa.

Agricultural Research and Extension Network (AgREN): run by the Overseas Development Institute, Portland House, Stag Place, London SW1E 5DP, UK, http:/www.oneworld.org/odi/. A network of interested individuals and organisations receiving twice yearly mailings with network papers. Free to libraries and those in the South.

CTA – Technical Centre for Agriculture and Rural Cooperation. PO Box 380, 6700 Wageningen, The Netherlands. Established to improve access to information on agricultural development in ACP countries. Produces magazine *Spore* and documentation services free to ACP countries.

International Institute for Environment and Development (IIED), 3 Endsleigh St, London WC1H 0DD, Tel: 44(0)171 388 2117; Fax: 44(0)171 388 2826; E-mail: iiedagri@gn.apc.org. *Participatory Learning and Action Notes*: produced by the Sustainable Agriculture Team every three months to enable practitioners using participatory methods to share their experiences. Free to non-OECD countries. Also Gatekeeper series of short briefing papers on policy issues and *Harmata* and *Issue Papers* on natural resource management in dryland areas.

LEISA Newsletter for low external input and sustainable agriculture. Institute of Low External Input Agriculture, Box 64, 3830 AB, Leusden, The Netherlands, Tel: 31(33) 494086; Fax: 31(33)4951779; E-mail: ileia@ileia.nl. Quarterly magazine on low external-input agriculture; can be requested free by organisations in the south.

Participatory Ecological Landuse Management Association (PELUM). PO Box MP1059, Mount Pleasant, Harare, Fax:/Tel: 744470; E-mail pelum@mail.pci.co.zw. A network of organisations in Southern and Eastern Africa collaborating on staff training and information sharing.

World Neighbours. 5116 Portland Avenue, Oklahoma City, OK 73112, US. Produces newsletter and other information for practical agricultural development; free to applicants from the South.

Appendix 3

References

Alexandratos/FAO (1995) *World Agriculture: Towards 2010* FAO, John Wiley, Chichester, UK

Arnaiz, M, Merill-Sands, D, Mukwende, B. (1995) *The Zimbabwe Farmers' Union: Its Current and Potential Role in Technology Development and Transfer* ODI AgREN Paper, ODI, London, UK

AZTREC (1995) *Impact Evaluation of AZTREC on Afforestation in Masvingo Province*, AZTREC, Masvingo, Zimbabwe

AZTREC (1996) *Annual Report*, AZTREC, Masvingo, Zimbabwe

Balderrama, S et al (1988) *Farming System Dynamics and Risk in a Low Potential Area: Chivi South* ICRA Bulletin 27, Wageningen, The Netherlands

BIDPA (Botswana Institute for Development Policy Analysis) (1996) *Study of Poverty and Poverty Alleviation in Botswana*, vol 2 & 3 (draft)

Bingham, S with Savory, A (1990) *Holistic Resource Management Workbook* Island Press, Washington, DC

Blackie, M (1990) 'Maize, food self-sufficiency and policy in East and Southern Africa' *Food Policy* 1990, vol 15(5) pp383–394

Bond, I (1993) *The Economics of Wildlife and Land Use in Zimbabwe* WWF, Harare

Central Statistics Office (1996) *Living Conditions in Namibia: Report from the 1993/94 Namibia Household Income and Expenditure Survey*, National Planning Commission, Windhoek

Cernea, M (1987) *Farmer Organisations and Institution Building for Sustainable Development*, Regional Development Dialogue 8

Chambers, R (1994) 'Participatory Rural Appraisal (PRA): The Origins and Practice of Participatory Rural Appraisal', *World Development*, vol 22(7)

Chambers, R (1997) *Whose Reality Counts: Putting the First Last* IT Publications, London

Chambers, R and Ghildyal, B (1985) *Agricultural Research for Resource Poor Farmers – The Farmer First and Last Model*, Agricultural Administration 20

Chambers, R, Parcey, A and Thrupp, L-A (1987) *Farmer First: Farmer Innovation and Agricultural Research* IT Publications, London

Comptroller and Auditor General (1993) *Value for Money Project on the Land Acquisition and Resettlement Project*, Government of Zimbabwe, Zimbabwe

Conroy, A (1993) *The Economics of Smallholder Maize Production in Malawi with Reference to Hybrid Seed and Fertilizer* unpublished PhD thesis, University of Manchester

Cousins, B (1995) *A Role for Common Property Institutions in Land Redistribution Programmes in South Africa* International Institute for Environment and Development, London

Cousins, T, Joaquim, E and Oettlé, N (1997) *Workshop Report: Pakkies Tenant Community* Department of Land Affairs, Pietermaritzburg

CURE (1996) *Assessment of Gender Issues in Community Participation in Natural Resource Management Activities* CURE, Chikawa District

Dobel, R (1996) *Participation and Institutional Arrangements: the Case of the Adaptive Research Planning Team (ARPT) in Zambia*, Research note in the Agricultural Research and Extension Network (AGREN), Newsletter 34, ODI, London

Dumont, R and Mottin, M (1979) *An alternative pattern of development for Zambia* GRZ/Institute Nacionale Agronomique, Lusaka

EDC (1996) *Literature Reviews and Fieldwork Plans: Agricultural Services Reform in Southern Africa Contract R6452CA* EDC Ltd, Hillside, Claypits La, Lypiatt, Stoud, GL6 7LU, UK

EDC (1997a): Whiteside, M *Encouraging Sustainable Smallholder Agriculture in Botswana* EDC Ltd, Hillside, Claypits La, Lypiatt, Stroud, GL6 7LU, UK

EDC (1997b): Whiteside, M and Carr, S *Encouraging Sustainable Smallholder Agriculture in Malawi* EDC Ltd, Hillside, Claypits La, Lypiatt, Stroud, GL6 7LU, UK

EDC (1997c): Copestake, J *Encouraging Sustainable Smallholder Agriculture in Zambia* EDC Ltd, Hillside, Claypits La, Lypiatt, Stroud, GL6 7LU, UK

EDC (1997d): Vigne, P and Whiteside, M *Encouraging Sustainable Smallholder Agriculture in Namibia* EDC Ltd, Hillside, Claypits La, Lypiatt, Stroud, GL6 7LU, UK

EDC (1998a): Whiteside, M *Encouraging Sustainable Smallholder Agriculture in Zimbabwe* EDC Ltd, Hillside, Claypits La, Lypiatt, Stroud, GL6 7LU, UK

EDC (1998b): Oettle, N, Fakir, S, Wentzel, W, Giddings, S and Whiteside, M *Encouraging Sustainable Smallholder Agriculture in South Africa* EDC Ltd, Hillside, Claypits La, Lypiatt, Stroud, GL6 7LU, UK

Eicher, C K (1995) 'Zimbabwe's Maize-Based Green Revolution: Preconditions for Replication' *World Development*, vol 23(5) pp805–15

Ellis-Jones, J and Mudhara, M (1995) in *Soil and Water Conservation for Smallholder Farmers in Zimbabwe: Proceedings of a Technical Workshop*, 3–7 April (1995), Masvingo

Elwell, H (1995) *An Assessment of the Performance of Minimum Tillage Practices in Zimbabwe in Terms of Drought Risk Alleviation, Yields and Cost Effectiveness* World Bank, unpublished

Fairhead, J and Leach, M (1995) 'Reading Forest History Backward: the interaction of policy and local land use in Guinea's forest savanna mosaic, 1893–1993' *Environment and History*, vol 1, pp55–91

Fairhead, J and Leach, M (forthcoming) 'Reframing Forest History: a radical reappraisal of the roles of people and climate in West African vegetation change' in G Chapman and D Driver (eds) *Timescales of Environmental Changes*, Routledge, London

FAO (1980) *Land Resources for Populations of the Future*, Report of the second FAO/UNFPA expert consultation, FAO, Rome

Farrington, J L and Martin, A (1987) 'Farmer Participation in Agricultural Research: a review of concepts and practices' *Agricultural Administration Unit Occasional Paper 9* ODI, London

Government of Botswana (1996) *National Development Plan 8* (draft)

Government of Malawi (1996) *Ministry of Finance: Fertilizer Discussion Paper*, Malawi

Government of Zimbabwe (1996) *Land Policy 5.2*, Zimbabwe

GRZ/CSO (1991) *Post-Harvest Sample Survey for 1990/91* CSO, Lusaka

GRZ/CSO (1994) *Post-Harvest Sample Survey for 1993/94* CSO, Lusaka

GRZ/FAO (1996b) *Review of the Agricultural Credit Management Programme in Zambia, 1994/95–1995/96: Constraints and Future Development Options*, FAO/MAFF Food Security Division, Market Liberalisation Impact Series no 16

GRZ/MAFF (1994) *The informal survey of Southern Province*, Farming Systems Research Team, Ministry of Agriculture, Food and Fisheries, Mt Makulu, Zambia

GRZ/MAFF (1991) *An Evaluation of the Agricultural Credit System in Zambia* Agricultural Credit Team, Zambia

Harvest Help (1994a) *Ndola Rural Technical and Smallholder Farmers' Association: agroforestry and natural resource management programme proposal*: internal report

Harvest Help (1994b) *Institute of Cultural Affairs of Zambia: Ipongo Development Programme Proposal*: internal report

Huntley et al (1990) *Southern African Environments in the 21st Century* Tafelberg, Cape Town

IFAD (1997) *Agribusiness Entrepreneur Network and Training Development Project (AGENT): Report 0762* Office of Evaluation and Studies, IFAD, Rome

ILEIA (1989) *Participatory Technology Development: A Selection of Publications* ILEIA, The Netherlands

Jackson, J and Collier, P (1988) *Incomes, Poverty and Food Security in the Communal Lands of Zimbabwe*, Occasional Paper, Department of Rural and Urban Planning, University of Zimbabwe

Kean, S A (1993) *The Institutional Politics of Agricultural Research in Zambia: a Model of Contingent Innovation*, PhD thesis, University of East Anglia

Land Tenure Commission (1994) *Report of the Commission of Inquiry into Appropriate Agricultural Land Tenure Systems*, Government of Zimbabwe, Harare

Lekgetha, J, Neill, C, Pearson, R with Kaunda, G and Kopulande, S (1995) *Collusion of Financial Disintermediation: An Analysis of Rural Financial Institutions in a Changing Economy*, Case study of Zambia, Consultancy report, publishers unknown

Machina, H (1996) *The Impact of Agricultural Liberalisation on Small Scale Maize Production in Zambia, 1975–90*, MSc dissertation, Department of Land and Economy, University of Cambridge

Magadza (1994) 'Climate Change: some likely multiple impacts in Southern Africa' *Food Policy*, vol 19(2)

Malawi Rural Finance Company: various annual reports

MAWRD (Ministry of Agriculture, Water and Rural Development, Namibia) (1997) *A Drought Policy for Namibia* initial draft

Mema, N and Murombezi, C (1996) *Evaluation of AZTREC*, AZTREC, Masvingo, Zimbabwe

Modise, S C (1996) 'Models of Wildlife Management: The Botswana Experience' *African Wildlife* Policy Consultation, ODA, London

Mombeshora, B, Mavedzenge, B, Mudhara, M, Chibudu, C, Chikura, C and Scoones, I (1995) *Coping with Risk and Uncertainty* ILEIA Newsletter, December (1995)

Moris, J (1991) *Extension Alternatives in Tropical Africa* ODI, London

Nkomesha, A, Russell, T D and Tunkanya, S (1993) *Farmers' Post-Project Evaluation of the Smallholders Development Project in Copperbelt Province* European Commission and Central Province Adaptive Research Planning Team

North, D (1990) *Institutions, Institutional Change and Economic Performance* Cambridge University Press, Cambridge

NRI (1996) *Report on the Scope for the Long Term Weather Forecasting to Affect Agricultural Planning in Southern Africa* Natural Resources Institute, Chatham

Overseas Development Administration (1988): Cusworth, J and Walker, J *Land Resettlement in Zimbabwe: a preliminary evaluation* ODA, UK

Overseas Development Administration (1996) *ODA Land Appraisal Mission to Zimbabwe* ODA, UK

Pantuliano, S and Whiteside, M (1997) Unpublished evaluation of ACORD programme in *ANG/S Support to Pastoralist Communities in Gambia* ACORD, London

Parsons, N (1982) *A New History of Southern Africa* Macmillan, London

Poulton, C, Dorward, A and Kydd, J (1998) 'The Revival of Smallholder Cash Crops in Africa Public and Private Roles in the Provision of Finance' *Journal of International Development*, vol 10:1, pp85–103

Pretty, J (1994) 'Alternative systems of enquiry for sustainable agriculture' *IDS Bulletin*, vol 25:2, IDS, University of Sussex, UK

Pretty, J (1995) *Regenerating Agriculture: Policies and Practice for Sustainability and Self-Reliance*, Earthscan, London

Pretty, J and Chambers, R (1994) *Towards a Learning Paradigm: New Professionalism and Institutions for Sustainable Agriculture* in *Beyond Farmer First*, IT Publications, London

RDSP (Rural Development Support Project for the Northern Communal Areas): Doughty, J (1997) *Notes and Planning Parameters for the Northern Communal Areas* (March 1997), unpublished draft

Rhoades, R and Booth, R (1982) 'Farmer-Back-to-Farmer: a Model for Generating Acceptable Agricultural Technology' *Agricultural Administration*, vol 11

Rukuni, M et al (1994) *Report of the Commission of Inquiry into Appropriate Land Tenure Systems* Government of Zimbabwe, Zimbabwe

SADC ELMS (1991) *Sustaining Our Common Future*, SADC, Maseru

SADC/ICRISAT (1993) *1992 Drought Relief Emergency Production of Pearl Millet Seed: Impact Assessment*, ICRISAT Southern and Eastern Africa Regional Working Paper

Saket, M (1994) *Report on the Updating of the Exploratory National Forest Inventory*, Ministry of Agriculture, Mozambique

SARDC/IUCN/SADC (1994) *State of the Environment in Southern Africa*, SARDC, Zimbabwe

Savory, A (1988) *Holistic Resource Management* Gilmour Publishing, Zimbabwe

Schwartz, L A (1994) *The Role of the Private Sector in Agricultural Extension: Economic Analysis and Case Studies*, Agricultural Administration (Research and Extension) Network Paper 48, ODI, London

Scoones, I (1994) (ed) *Living with Uncertainty. New Directions for Pastoral Development in Africa*, IT Publications, London

Scoones, I (1996) *Hazards and Opportunities: Farming Livelihoods in Dryland Africa, Lessons from Zimbabwe* Zed Books, London

Sikana, P (1994) 'Alternatives to current research and extension systems: village research groups in Zambia' in Scoones (1994)

Stringfellow, R (1996) *Smallholder Outgrower Schemes in Zambia* NRI, UK

Thomas, S (1995a) *The Legacy of Dualism in Decision Making within CAMPFIRE* IIED, London

Thomas, S (1995b) *Share and Share Alike? Equity in CAMPFIRE* IIED, London

Tiffen, M, Mortimore, M and Gichuki, F (1994) *More People, Less Erosion: Environmental Recovery in Kenya* John Wiley, Chichester

Tyson, P (1987) *Climate Change and Variability in Southern Africa* OUP, Cape Town

UNDP (1994) *Human Development Report* UNDP, Oxford

UNDP (1996) *Human Development Report* UNDP, Oxford

Walling, D 1984 'The sediment yield of African rivers' in *Challenges in African Hydrology* Institute of Hydrology, Wallingford, UK

Wellard, K and Copestake, J (1993) *Non-Governmental Organisations and the State in Africa: Rethinking Roles in Sustainable Agricultural Development* Routledge, London

Whiteside, M (1996a) *Perspectives on the Effects of Conflict at the Village Level – Mozambique*, Occasional Paper Series on the Environment and Development in an Age of Transition, University of Leeds, UK

Whiteside, M (1997) *Two Cheers for RRA*, Notes on Participatory Learning and Action 28, IIED, UK

Wiggins, S (1995) 'Change in African Farming Systems Between the Mid 1970s and the Mid 1980s' *Journal of International Development*, vol 7:6

Wood, A, Kean, S, Milimo, J and Warren, D (1990) (eds) *The Dynamics of Agricultural Policy and Reform in Zambia* Iowa State University Press, Iowa

Woodhouse, P (1991) *Participatory Agricultural Research and Resource Management: What Role for Extension Services in Africa* Paper presented to conference on Sustainable Agricultural Development, University of Bradford, Bradford

World Bank (1992) *Zambia Agriculture Sector Strategy: Issues and Options*, Report no 9517-ZA

World Bank (1996a) *Zambia Agricultural Sector Investment Programme: Staff Appraisal Report Southern Africa* Department of Agriculture and Environment Division

World Bank (1996b) *Prospects for Sustainable Growth in Zambia, 1995 to 2005* Macro Industry Division, Southern Africa Department

Zambian Farmer (various) Monthly newspaper of Zambia National Farmers Union (ZNFU)

Index

Printed in the United States
by Baker & Taylor Publisher Services